WHY MALES EXIST

An Inquiry into the Evolution of Sex

WHY MALES EXIST

AN INQUIRY INTO THE EVOLUTION OF SEX

by FRED HAPGOOD

WILLIAM MORROW AND COMPANY, INC.
NEW YORK 1979

Library of Congress Cataloging in Publication Data

Hapgood, Fred.
Why males exist.

Bibliography: p.
Includes index.
1. Sexual behavior in animals. 2. Reproduction. 3. Evolution.
I. Title.
QL761.H36 591.5′6 79-13805
ISBN 0-688-03546-9

Printed in the United States of America.

First Edition

1 2 3 4 5 6 7 8 9 10

To the fathers of ethology:

Julian Huxley, Karl von Frisch,

Konrad Lorenz, and Nino Tinbergen

ACKNOWLEDGMENTS

MY GREATEST DEBT IS OWED TO THE ETHOLOGISTS WHO brought back the stories that light up these pages. Unfortunately, they are far too numerous to mention individually; even my bibliography does not do them justice. The professional literature is amazingly rich. I cannot count the times I opened a journal in search of a specific reference and found myself reading one article after another, each more wondrous and delightful than the last. I tried to convey some feeling for these reports by quoting from them frequently.

A second debt is owed to certain specific theoreticians and scholars. Preeminent among these is Ernst Mayr, whose magisterial book, *Animal Species and Evolution*, gave me more instruction and excitement than any other book I can recall. E. O. Wilson's *Sociobiology* first revealed to me the degree of skill and intelligence with which animals live their lives. George Williams, in his *Sex and Evolution*, challenged me to think about these problems, in large part by the style of his example, which was one of great openness, honesty, and modesty. Robert Trivers was very important in focusing and supporting my energies on this subject. It is impossible to imagine this book without the famous contributions that his imagination, incisiveness and intellectual stubbornness have made to behavioral biology. Michael Ghiselin's *The Economy of Nature* and *The Evolution of Sex* were also extremely instructive.

Finally, I wish to thank John Hartung, Frank Benford, John Weinrich, Sandi Friedman, David Crews, Richard Wrangham, Mrs. Tamsin H. Hapgood, Sara Dunlap, Steve Warshall, Mrs. Herbert P. Mayo, Vin McLellan, and Sarah Hrdy for reading and commenting on the manuscript at various stages.

CONTENTS

Acknowledgments *7*

Introduction *11*

I. The Extravagance of Males *15*

II. The Rule of the Game *20*

III. Living Sexlessly *25*

IV. An Overview of Some of the Cons and Pros of Sex *32*

V. The Transition to Sex *35*

VI. The Consequences of Being Sexual *53*

VII. The Evolution of Gender *67*

VIII. The Prisoner of Gender—an Unresolved Mystery *78*

IX. Doing the Paperwork and Other Issues *84*

X. Male Society—the Society of Competition *98*

XI. Female Influences on Male Society *110*

XII. The Cost of Male Competition to Females *128*

XIII. Males With an Edge *139*

XIV. Male Leverage and Equality *149*

Notes *189*

Index *201*

About the Author *213*

INTRODUCTION

THIS BOOK HAS TWO AMBITIONS: ONE EXPLICIT, ONE IMPLICIT. The explicit intention is to present a problem that few people know even exists, and that is the mystery of males, why they're here and what they're for. The dilemma is best stated with an illustration. Grizzly bear females carry, nurse, protect, and train their young. Male grizzlies take no role in any of these activities; they make no obvious contribution at all to the energy needs of reproduction. All those burdens, such as having to hunt and forage for a whole family instead of just for one, fall onto the female's shoulders.

Given that fact—and this arrangement is very common in nature—what good are males at all? Why did they arise, or evolve, in the first place? There are many species that get along perfectly well without them. Some of these reproduce without using sex (that is, gene-mixing, or fertilization) at all; in others, all the individuals make both sperm and eggs and then reciprocally fertilize each other. Why are these alternatives to males not more widely pursued? Why aren't we all bisexual, with one gonad acting as a testes and one as an ovary (only one of each is necessary for either reproductive role anyway)?

Perhaps there is no good answer to this question. Males may well be an evolutionary dead-end that will be abandoned when another system, less parasitical on and wasteful of female energies, appears. But there are other possibilities, and these can be summarized by saying that males might make sense if they allow the needs and desires of females to control their actions.

The second, implicit, intention is to show that animals and

humans belong to the same culture. We humans are so flexible in our behaviors, we are adapted to so many different environments, that there are very few species anywhere on the globe that are not dealing with issues that are familiar and important to us. To put the matter in its most egoistic form, human behavior summarizes that of all the members of the animal kingdom. No doubt these few words have failed to define this idea with any clarity, but there is no point in continuing to press it in general terms. There is a text at hand, with lots of examples, and the best definition is a good illustration.

Nevertheless all living things have much in common.

—CHARLES DARWIN

CHAPTER

I

THE EXTRAVAGANCE OF MALES

WHAT IS THE DIFFERENCE BETWEEN THE LIVING AND THE dead? between biological creatures, some as immobile as rock and as passive as windblown sand, and inorganic substances and reactions, some of which, like fire, can grow and leap as responsively as most living things? This is an old question, with many answers, but the answer that seems to work best is: Living things can reproduce themselves. Dead material loses its patterns with time; it erodes, crumbles, decays, dissolves. Living organisms can preserve their forms intact by building them anew, over and over, and so pass themselves down through time. Life is the power of changing substance while preserving form.

It is as awe-inspiring, as overwhelming, as any of the other natural powers. Our planet is heavy with forms straining to stamp out copies of themselves, most of them trying to transform the flesh of other life-forms into their own patterns. Some naturalists celebrate this rank, riotous fecundity; others detect something driven and horrible at its heart. "In this repetition of individuals is a mindless stutter, an imbecilic fixedness . . . ," Annie Dillard, one of our best contemporary nature writers, has written. We are concerned, in these pages, with one small aspect of this fierce power, and that is the fact, and the implications of the fact, that all this abundance (with trivial exceptions) is the work of females.

In other words, among species with genders, it is the females that have control over the reproductive cycle; the males appear to be virtually irrelevant. The single-parent family is the rule in nature and that parent is almost always female. It is the females that watch the environment to clock the arrival of favorable conditions, secure the food to make and nourish her eggs, find and/or prepare a proper place to lay her eggs or leave her young, and incubate, nurse or feed, train, and protect them. By and large males take no part in these activities.

There is only one wide-spread class of exceptions to this generalization; perhaps in 90 percent of the bird species the male shoulders a fair share of the reproductive responsibilities—helping to build the nest, defend the site against competitors and predators, and feed the young. In no other class of organisms does the percent of species with working males rise higher than 10 percent, and usually it is far less than that. To draw a parallel, if females and males were partners in a construction project, females would have to be imagined as putting up the capital, hiring the subcontractors, acquiring the necessary materials, arranging for the permits, and devising half the design changes. The males contribute half the design changes and collect 50 percent of the profits—the same share as the females.

This parallel is not quite fair to males; they cannot, in fact, afford to loll about. Since males have only a peripheral role in the reproductive cycle, it takes them virtually no time at all to reproduce—all they have to do is mate. That means that, at least theoretically, one male might be able to fertilize all the females in a large area; an achievement that would shut all the other local males out of reproducing altogether. Since, as Darwin taught, natural selection chooses individuals that reproduce better, one would expect evolution to produce males who would go to great lengths to avoid being so excluded. And so it has, with the result that, at least during breeding season, being a male is an exhausting business that involves competing continuously with many other males.

Sometimes this competition takes the form of a direct physical attack by one male on another; most often it means noise, a great deal of noise, produced nonstop, hour after hour, sometimes day after day. These are the sounds of music with which the hills are alive—the choruses of frogs; the shrieking of the cicadas; the staccato calls of crickets; the repertories of male songbirds; the yowling of tomcats on the prowl. And from the bellowing of bull stags to the peeping of tree frogs all this sound serves the same ends; it intimidates rivals and attracts and/or persuades females. It is all huckstering and swagger matches; winning and losing; an endless game of who's better.

So the fact that males are excluded from any real participation is reproduction does not mean that they have it easy; if anything, their lives are harsher and more taxing than those of females. A biologist named Valerius Geist has measured the skulls of mountain sheep and found that males with larger horns died younger. He believes that the large horns bring males sexual success, but that that success has a cost. The sheep have to constantly ward off threats by other males to "their" females. This destroys the fat reserves they need to see them through the winter, and therefore sexually successful males die earlier. Geist writes:

> Once rams have entered their period of dominance, of high social success, of guarding and breeding with females, they begin to age and die rapidly. Mortality increases five- to eightfold over that of younger, subordinate rams. Once they have become breeding rams, the males have not long to live.

Sex can be shown to be bad for males in other ways as well. There are: spiders that capture male insects by giving off the odor of a receptive female; firefly predators that catch males by giving off the distinctive flash pattern of a female firefly; cricket-hunting wasps that are attracted by the song of male crickets. When wild dogs hunt gazelles they do so in a specific formation that makes it easy to catch any gazelle that tries to double back. They do this because often the gazelle they are pursuing is a male whom they have chased from his mating territory, and who will be desperate to twist around and return to it. When he turns he is caught. To be blunt about it, sex makes males do very stupid things; they seem to throw all concern for survival to the wind. And there is a perverse logic to it all. Any male that tried to live a sober, sensible, careful life might live for a long time, but he would never reproduce. His rivals, the reckless bravos, might lead short lives, but they would be full ones, since—and this is the key point—male reproduction is such a quick affair that even a short life can be adequate to mate with many females.

This looks like a totally pointless system. Everybody seems to end up worse off. Females are denied the benefits of male assistance; males, because they have nothing useful to do, are forced to blow their energies off into the atmosphere and spend their time fighting with each other. The male striped bowerbird of New Guinea builds, as his form of sexual advertisement, a display hut. First he selects a plant with a strong central shoot and strips it down. Then he weaves a thick column of fibers around the shoot and decorates it with a profusion of flowers, berries, colorful leaves, and iridescent beetle wings. Next he builds a two-by-three-foot roof over the column, leaving a portal through which the column can be admired, and lays down a lawn of moss in front of the entrance, scattering red fruit, flowers, and berries over it. Finally he circles the entire structure with a low wall.

This is a lovely tale, but what is the point? Why devote so much time and energy to these exotic and sumptuous love chambers? Is the bowerbird not in danger of being rendered extinct by some other, more puritan species that will put the same time and energy into feeding more young? The common American grasshopper has a courtship dance with eighteen separate patterns in it, making it considerably more complicated than the disco hustle. Why is it necessary?

When one looks at females in nature one sees industry,

progenitiveness, and efficiency; when one looks at males one sees the most amazingly elaborate forms of wastefulness. How is it possible that both genders evolved together, as they must have? And are males really what they seem to be, an evolutionary frivolity, an extravagance with no practical point to them at all?

CHAPTER

II

THE RULE OF THE GAME

OVER A HUNDRED YEARS AGO CHARLES DARWIN INVENTED the way we now answer questions such as "Why do males exist?" or, for that matter, "Why do fingernails exist?" He decided to ignore one part of the question, which was exactly why the first male, or first fingernail, appeared—and to concentrate instead on why males or fingernails came to be so common. He just assumed that there is so much noise in the reproductive process that slightly offbeat and novel ways of going about life will be continuously and spontaneously thrown off. What interested him was the next step; how these novelties became standard equipment; how they spread from the very first individual to every member of the species.

Darwin reasoned that the only way in which a trait could succeed in thus spreading itself to larger and larger numbers would be if it met two conditions. It had to be inheritable, and it had to enhance the fecundity of its bearers. If it met these two conditions then it would be passed on to larger numbers of individuals with each successive generation, until eventually it became standard equipment.

That is Darwinism, and in its essentials it has changed very little since 1859, when *The Origin of Species* was published. (In fact in some respects contemporary evolutionary theory is closer to Darwin than mainstream evolutionary thought was twenty years ago.) One key point that should be underlined is that the theory is explicitly competitive. Its basic model is that of a race between two individuals who are in all respects the same—they have the same genes and live in the same environment—except that one has the novelty and the other doesn't. A Darwinian explanation of an existing trait has been achieved when one has devised a plausible argument for believing that all else being equal, creatures with the novelty will be more fecund than those without it.

For example there is a fairly large group of tropical fish, called "mouthbreeders," who incubate their eggs in their mouths after the eggs have been laid—supposedly to protect them from predators. Sometimes the male is the incubator (the mouthbreeders constitute a major fraction of the tiny percent of fish species [less than 5 percent] in which the males contribute some work to the reproductive cycle) and sometimes the female. In the latter cases the female lays the eggs, takes them into her mouth, and swims off with them still unfertilized. The males of at least one species of female-incubating mouthbreeders (*Haplochromis burtoni*) have egglike designs on their anal fins; they also have the habit of swimming down to the bottom of the lake and flicking these fins. Why do they have these designs? And why do they behave in quite that way? There is one additional fact about these fish that instantly suggests a classically Darwinian explanation. When a female notices a male flicking his fins on the bottom of the lake, she immediately swims down to him and opens her mouth, just as though she were trying to pick up the eggs pictured on the fin. What she gets instead is a mouthful of sperm.

So if one imagines the possession of both the egg designs and the knowledge of how to use them to be a single inheritable trait, and one imagines males with this trait competing for females with males that lack it, it seems easy to see why that trait would spread. The males with the designs have hit upon a trick that allows them to get more females, and therefore produce more progeny, at the expense of the other designless males.

There are two common misunderstandings of evolution that should be mentioned explicitly. The first is that evolution promotes survival. This is only indirectly true. What evolution promotes is reproduction. Survival is obviously important to reproduction, but if and when procreativity is enhanced by a shorter life-span, or by a life-style that necessarily involves a shorter life, then early death will be selected. In the last chapter there were several examples of males leading risky, exhausting, stressful lives that probably killed them much sooner than if they had chosen a more serene life-style. Nonetheless, given the high levels of male-male competition that exist in nature, only males that expend their energies in savage bursts, without counting the cost, can hope to reproduce at all. Sometimes a shorter life-span might even be selected directly. There are two antipredator strategies generally adopted by butterflies and moths. One is to secrete poisonous toxins; the other is to look like a poisonous insect, but without actually producing toxins. The genuinely poisonous insects tend to have a longer postreproductive life-span than the mimics, who often die immediately after laying their eggs. One explanation of this difference is that the poisonous insects live longer so as to be available for the education of the local birds—once they eat the parents, they'll steer clear of the offspring. The mimics, contrariwise, die quickly to prevent the local birds from catching on! Whether or not this is true I do not know, but the example makes the point. Survival is only important to the degree that it enhances reproduction.

The second common misunderstanding is that the survival of the species counts for something in evolution. If so, that fact can't be accounted for by Darwinian theory. Darwinian evolution always works by, through, and for individuals, improving their reproductive efficiencies for reasons that have to do with their own local circumstances, not because of any problems that might face the species as a whole. If the species survives, that is well and good, but it does so as the by-product of a process that only recognizes individuals (or, at most, families). Darwinism is surely not the last word on this subject, and no doubt on some issues, at some level, interspecies competition is important, but for the purposes of the problems being discussed on these pages, evolution is assumed to be a process that goes on between individuals of the same species.

So the rule of the game—the game itself is to make some sense out of the existence of males—is that one has to imagine a non-helpful masculinity evolving by competing against a system in which EVERYONE works to make offspring—and winning! Further, because sex is a relationship involving two creatures, a really satisfactory explanation would be one that showed how masculinity could win against this competitive system in two different contests. The first would show why half the species found it worthwhile to give up making offspring and become males; the second contest would explain why the other half of the species, those that stayed female, tolerated this development. Why did they get into male-making at all? Assuming that the females had an alternative, so that they could have turned out a full set of offspring all of which were industrious, efficient, and productive, why should they have devoted half their output instead to making creatures that make noise and fight? According to Darwin there must have been a reason why both those creatures that became males and those that didn't became better, more procreative organisms as a result. It is certainly not intuitively obvious why this should be so.

Fortunately evolutionary theory also suggests a way to begin untangling these mysteries. If evolution is a competition, then both competitors are equally important to the outcome of the race. Masculinity did not evolve in a vacuum but against a specific alternative; perhaps the answer is not that masculinity has anything especially positive to recommend it —it's just that the alternative is so much worse. There are many species that live without males on the planet, and while the fact that they do live genderlessly and/or sexlessly shows that in their environments males are unnecessary, one can still hope that they might give some indirect clue as to what might be the problem with trying to live without males in our world.

CHAPTER

LIVING SEXLESSLY

VERY FEW CLASSES OF CREATURES ARE AS RIGIDLY COMmitted to a single mode of reproduction as we vertebrates are. Most organisms (though not most of those we can see easily) reproduce asexually much of the time, but retain the option of switching to gene-mixing reproduction once every ten, or hundred, or thousand generations. In other words, it is possible to arrange all the forms of life along a sexual spectrum according to how frequently they reproduce sexually as opposed to simply cloning their progeny. If we did this, warm-blooded vertebrates would be on one extreme and bacteria on the other.

Bacteria are the starting point, a reference point, for a good deal of thinking in biology. They are the smallest class of creature, and so thinking about the role of size—or lack of it—naturally centers on them.* They reproduce fastest, are the most fecund organisms, and so provide a standard of success that is of special interest to Darwinians. If any life-form can be called the most successful, important, and best-adapted on the planet, it can only be bacteria. They are the dominant life-form, making up maybe a quarter of the biomass. Probably 99 percent of all the biological individuals on Earth are bacterial.

* Viruses are smaller, of course, but for the purposes of this chapter bacteria and viruses can be treated as a single group.

Until recently biologists believed that they came only in three shapes: rods, spirals, and spheres. We now know they can come in any shape: as threads, rings, knots, even as stars. There are forms that live without any shape, like amoebas. Some bacteria are the only creatures to have a biological propeller, a real rotor that whirls around, sticking out of their backsides. Other bacteria glide like snails over self-generated lubricants; others tether themselves with a ribbon of gum to a surface and then pay it out behind them as they float, like tiny kites, out into the nutrient stream; others root themselves on rigid stalks, like grains of wheat; while others form vast colonial fishing nets out of their bodies in which they capture gases and snare creatures hundreds of times larger than they. The largest bacterium is more than a thousand times the size of the smallest—a greater range than that between mice and whales. They can live and flourish under the most implausible extremes of heat, cold, pressure, aridity, salinity, and nutrient deprivation, and are always the first life-form to appear as environments grow hospitable and the last to vanish.

They are the most important because every other class of living thing is dependent on dozens of bacterial processes. Bacteria help weather rock to make soil and then stock it with nutrients they produce themselves. All plants, for example, depend entirely on bacteria to process and refine atmospheric nitrogen so that plants can use it. Bacteria form the first link, or, perhaps better, the anchor, of the food chain. (They are the only creatures that could live on an otherwise sterile planet, independently of any other life-form.) And they are the recycling engines of Earth, without whose efforts the nitrogen, carbon, sulfur, and iron cycles would all collapse. From an ecological point of view we vertebrates live like cockroaches and silverfish, in the corners and interstices of a great house built for and maintained by another tenant entirely.

Finally, they are the best-adapted and most highly evolved life-form in the simple sense of having been at it longest. The first five-sixths of the history of life, three out of three and a half billion years, was given over to bacterial evolution. Their kingdom has within it a number of members that have been continuously viable for billions of years.* It has been hard for us to recognize the evolutionary position of bacteria because we look at life with a Western, technocratic, chrome-and-tailfins theory of evolution in which evolutionary progress is calculated by the number of different traits and characters a species has managed to accumulate. Simplicity, clarity, and economy count for nothing. If it did (and perhaps someday, when we devise an Eastern theory of evolution, it will), then bacteria would be recognized as the supreme achievement of terrestrial evolution, since which the process has deteriorated into designing one Rube Goldberg contraption after another.

The most important feature of the bacterial life-style is its boom-and-bust quality. In general bacterial populations either are growing explosively, collapsing catastrophically, or lying completely dormant. They live like Dionysiacs, waking only when they can live orgiastically, gulping food with a technique that wastes far more than it uses, reproducing riotously, and polluting everything about them with their toxins. As soon as they have destroyed the conditions that woke them up, they just slip away again into dormancy. Bacteria are total hedonists, living only for the short run (and the shortest of short runs, at that). And it is this life-style which makes possible every one of the scrimp-and-save, puritan, accumulating species on Earth.

Being small, the way bacteria are, tends to increase dependency on local environments. Small creatures generally (compared with larger ones) can cover less ground and must find what they need in a smaller area. Their internal volume is less, and so they are not as able to store reserves, to build up a bank of nutrients that would allow them to cross, or survive, a depleted environment. In general, they must rely more on waiting for the right conditions, all together and at

* Instead of Linnaeus's two-kingdom plant-animal system I use a five-kingdom organization now entering the texts: (1) bacteria and viruses; (2) protozoa and algae (the protista); (3) fungi; (4) plants; and (5) animals.

once, to come to them. This is the strategy pursued by many forest flowers, for example, which can wait in the form of seeds for decades before a tree blows down and a patch of sunlight appears on the forest floor. Bacteria also have the knack of waiting patiently, in a low-energy state, often as spores, for the moment when everything is just right for explosive growth. They can endure the most extreme insults when in this dormant period—boiling, freezing, high levels of radiation, corrosive chemicals, long periods of time (sometimes years)—and still be able to respond to the return of their own kind of conditions. (There is a theory that life was first brought to Earth as an infection by extraterrestrial bacterial spores so long-lived and hardy that they were able to survive the journey from another solar system.) It is the arrival of these conditions that controls the turnover of generations; a bacterium might wait weeks or months for some nutrient to appear, during which time it would not reproduce at all. When that nutrient does appear, it might reproduce every twenty minutes. In human terms this would be like waiting for ten thousand years so that everything would be just right in order to have one child—and then having hundreds of them.

Small size is both an advantage and a disadvantage in competing with other organisms for a food source. On the one hand, larger creatures can just brush the smaller aside; on the other, it takes less material to make a smaller organism. A given quantity of material makes more little animals than big ones, and little ones don't need to grow as long, or as much, to get to their adult, reproductive, weight. So the strategy that small organisms pursue is rapid, pell-mell reproduction.

As a rule these reproductions are asexual. First an individual copies its genes and then distributes one copy to each end of itself. Next it splits across the middle, making two little bacteria, which suck up food from the area, fatten to adult size, and split again. Under the right conditions some bacteria can do this every ten minutes, and a great many species can knock out a new generation every thirty minutes. The gene-copying itself is done just the way you might make a copy of a tape recording. Bacterial genes are organized in a long string—usually five to six hundred times the organism's body length—with joined ends. This string is stacked in the cell so that it looks like the cooling coils in the back of a refrigerator. When it is copied, the whole length rotates, very quickly, past a stationary copying head that throws off a duplicate in the opposite direction. When bacteria do reproduce sexually, one individual extends a tube into a second down which this duplicate gene tape then slides. Some fraction of the introduced string—seldom the whole length—is then incorporated into the genetic coil of the recipient. (In other words, the bacterium that accepts the genes is fundamentally changed by the experience; its genetic identity is permanently altered. As an example, a bacterium that lacks the genes to make a tube and act as a gene donor can sometimes do so after being mated with. This has led some bacteriologists to say that bacterial sex is a venereal disease, because it is spread by copulation. Most sexual creatures only mix their genes at arm's length, off in their progeny, while keeping their own personal gene sets out of the process.)

The important thing about bacterial sex though, is that (compared with larger creatures) it is probably very rare. Scientists would like it to be more common, since many kinds of experiments become possible if bacteria can be persuaded to mate in a controlled way. Some progress has been made in this field, but there are still many kinds of bacteria that no one has succeeded in inducing to mate. Of course what goes on in a test tube may have nothing to do with events in the wild. Maybe out there bacteria mate all the time; the spread of antibiotic resistance shows that some genes do get passed around. But most bacteriologists believe that the average bacterium is extremely unlikely to have a sexual experience. For one thing, bacteria have many adaptations that seem designed to make sure that sex is kept to just a marginal role in their lives. Usually just a little bit of the gene string is accepted; the genes that promote transfer are sometimes not passed on, and there are other adaptations of the same kind.

So one could argue that the most successful, important, and highly adapted form of life is also the one that relies on sex the least. Given all the obvious disadvantages of sex—its rigidity, complexity, and the riskiness of gene-mixing—this should not be totally unexpected. Surely the minimal inference one can draw from the fact that a sexless life-form is the dominant organism on the planet is that doing without sex, or at most doing with very little, can be a very good thing. The converse of that observation is that organisms getting involved with sex are shouldering a burden. We ought not expect them to do so for frivolous or insubstantial reasons.

At the beginning of this chapter it was remarked that it is possible to arrange all the forms of life along a spectrum, according to how often they reproduce sexually. Bacteria lie on one end of this spectrum; warm-blooded vertebrates on the other. In between these two there are a great many organisms that dabble in both modes. In the next chapter we will look at some of these, with the hope that catching them at the moment when they shift into sexual reproduction will also show us what their reasons for doing so might be.

A GLOSSARY OF REPRODUCTIVE TERMS

sexual reproduction—a reproductive mode in which two organisms combine their sex cells, creating a set of offspring in whom the genetic identities of both parents are mixed.

asexual reproduction—the creation by one parent of a set of offspring that is genetically identical both to the parent, and to each other. Examples of asexual reproduction include the splitting of bacteria, or amoebas, the division of cells in our own bodies, and the reproduction of vines through creepers. Cloning is asexual reproduction.

gender—a form of sexual reproduction in which the species divides into two symbiotic specializations—one which donates a sex cell, and the other which accepts it. In common parlance, males donate sperm; females make eggs and accept sperm.

bisexuality—a form of sexual reproduction in which sexual specializations either do not form or are greatly reduced in importance. In these species the organisms can make both sperm and eggs, either at the same time or alternately, and can therefore be both male and female during the same life-span. Examples include many plants and invertebrates. Many species of snails and worms are bisexual.

CHAPTER

IV

AN OVERVIEW OF SOME OF THE CONS AND PROS OF SEX

THERE ARE THREE ALTERNATIVES TO WASTEFUL MALES: males, like beavers or nesting birds, that work; bisexuals; and creatures that reproduce asexually. The key feature of these three systems is that the organisms in them clearly channel their energies in useful directions. Of them all the most radical is asexuality, or sexlessness, in which the progeny are simply cloned. Sexlessness is reproduction without gene transfer; it not only abolishes wasteful males but all males, productive or not. So the most sweepingly effective way of doing without males is to learn to do without sex and reproduce as a single parent.

Taking this route solves a number of other problems at the same time. Leaving males to one side, sex itself, gene mixing, is a problematic and troublesome business. It burns up energy, takes time, requires a complicated and by no means trouble-free set of physiological aids, and often increases vulnerability. Sex can also be limiting. If reproduction requires a partner, then creatures who find themselves in favorable habitats, but alone, will not be able to colonize those environments.

Finally, since sexual creatures are constantly mixing and blending their genes from mating to mating, generation to generation, they reproduce themselves only in a general, low-fidelity, sense. The story is told of a famous actress writing to George Bernard Shaw and suggesting that they conceive a child, for, she said, with her beauty and his brains, what a paragon the infant would be! Shaw turned her down, suggesting that their child might as well inherit his beauty and the actress's brains. His point was true enough; but neither possibility was anything as likely as that their child would have had neither, but been much closer to the norm in both respects. As a biological fact, neither Shaw nor the actress could reproduce themselves except as vague approximations. This is because humans are sexual animals. Such a behavior seems completely un-Darwinian. Each parent is already a winner in the struggle for existence; why should it be selected to turn its back on the genetic constitution that proved so serviceable in its personal experience? Of course the parent is mating with another winner, but surely breaking up a gene complex that has already worked well once and mixing it up with the genes of a second is a real gamble, full of imponderables.

Most creatures appear to have found one or another of these reasons to avoid sex persuasive, since only a relative handful engage in it consistently. It appears otherwise to us, since if we were to rank all organisms by whether they were readily visible to the unaided eye we would also have separated the sexual from the asexual, or the drastically less-sexual. As a rough rule, organisms that cannot be seen —viruses and bacteria, the larger one-celled organisms like the amoebas, and the smaller multicellulars—reproduce sexually much less often than creatures that can be seen. The reason that only a handful of the reproductive events in the world are sexual is that the smaller creatures are both incomparably more numerous and reproduce much more often.

Why organisms reproduce sexually has been argued for decades. Some believe that sex is an advantage because it speeds evolution; others, because it slows evolution down, preventing too tight an adaptation to local, transient circumstances. A third theory is that sex repairs and rejuvenates the genes. A fourth theory points out that sexually produced life-forms have two complete and (usually) different sets of genes, one from each parent; the virtue of sex is thus the versatility and resilience that these "backup" genes give. A fifth idea, and the one relied on here, is that sex is selected when it becomes important for an individual to leave behind a set of offspring with a wider range of specialized skills, a richer variety of vocational orientations, than she could possibly make on her own.

Untangling the purpose of sex is obviously not a simple business, but it has an intimate connection to our subject, which is males, and how they happen to be here despite their apparent wastefulness. It is striking that (a) we don't understand what (if any) function males perform; (b) sex makes males possible, that is, it was the evolution of sexual reproduction that opened the way for the evolution of males; and (c) the purpose of sexual reproduction is itself somewhat mysterious. Given two such closely related problems one might hope that they can explain each other in some way, so that if we really understood sex then we would see the point of males as well. Intuitively we all make this connection between genders and gene transfer, since we use the word "sex" to apply to both traits. Intuition isn't necessarily right, of course, but the possibilities it suggests are as good as any with which to begin.

CHAPTER

V

THE TRANSITION TO SEX

IN MOVING TO THE NEXT STEP, OR STOP, ALONG THE SEXUAL spectrum, we cross what many biologists feel is the single most unambiguous and profound distinction that can be made among living things—that between bacteria on one side and protozoa and algae on the other.* All these forms are still single-celled, and to a person without a microscope the distinctions can mean nothing, since none of them will be visible. But someone with a microscope wouldn't know how to begin listing the differences.

One is that the protists, as the members of this kingdom are called, are mostly several hundred times larger than bacteria. A protist that feeds on bacteria, as many do, has about the same relationship to them as does a human to sausages or french fries. But the most important set of differences involve degrees of internal complexity. Internal organs appear for the first time in the protista, including digestive and excretory systems. Most have specialized energy organs that extract and store energy from the sun or from chemical reactions. They can store materials inside their bodies as well: fats and starches and excretion deposits. They have attack and defense mechanisms, including paralyzing secretions, ex-

* Well-known examples of protozoa include the amoebas and the paramecia.

plosive darts, and spines, and powerful propulsive devices, which, together with their heavier body mass, allows them to swim more smoothly, forcefully, and effectively than bacteria ever could.

The increase in complexity is echoed (and emphasized) by an order-of-magnitude increase in the amount of DNA contained in protista genes. Some orders of algae have as much DNA as mammals. (It is thought that the way in which evolution made the "jump" from bacteria to protista was by several different kinds of bacteria forming a single, symbiotic organization. Over time the members of this association became so harmoniously integrated, their identities as individuals so commingled, that they evolved into the organs of a larger order of creature, and so evolved the creature as well.)

These larger, more complicated organisms rely on sex much more than bacteria do—though they still reproduce asexually more often than not—and their sex is more thoroughgoing, involving a process by which the entire genetic libraries of both mates get completely shuffled together. (As opposed to just a fragment of the genes, as in bacteria.) That said, we have just about come to an end of all the possible useful generalizations one can make about protista sex. It is astonishingly various. Perhaps the best one can do is to give a few examples that suggest this diversity.

A first example (any one of thousands would do as well) might be the case of *Trichonympha* (trik-uh-NIMF-uh), a pear-shaped protozoan that lives in the gut of cockroaches, where it eats scraps of cellulose. Hundreds of flagella are attached to the narrower part, or head; these drive the creature around by pulling it, breast-stroke style, head-first, through the medium. When it finds a piece of cellulose, a pseudopod reaches out of the more bulbous posterior and pulls the scrap into the organism.

Usually a *Trichonympha* reproduces asexually, by slitting itself from top to bottom. When it detects the cockroach's molting hormone, however, it switches to sexual reproduction. (When cockroaches molt they also shed their gut lining. All the microflora that live in the gut are ejected with the lining, so that to the *Trichonympha* the sudden appearance of the molting hormone is the same thing as receiving an eviction notice.) First it makes two copies of its genes and distributes them to either side, seals itself up into a cyst, and splits vertically into two organisms. So far this is the same procedure as that followed during asexual reproduction. But this time when the two progeny squeeze out of the cyst back into the cockroach gut they are not identical—one is perceptibly smaller than the other. (This smaller one is sometimes called the "male" because of its subsequent behavior.) The same process goes on with all the other *Trichonymphae* in the insect's gut, and soon the entire tract is pulsing with these special, differentiated, progeny. Soon a mass copulation, described by the German zoologist, Wolfgang Wickler, begins:

> . . . the male individual follows the female and penetrates her from behind through a special zone of plasma. . . . [In the terms used here, through the bulbous end.] A typical *Trichonympha* female is recognizable by the number of little dark pigment spots arranged in a dense ring on her rear cell section. The male, by contrast, has only a few of these little dots distributed freely over the entire body of the cell. But there are all manner of transitions between the two extremes, and one can determine how strongly developed the female tendency of such a unicellular individual is by the number of dark dots. During a typical copulation, an individual with only a few dots penetrates one with a dense ring of dots . . . so it is playing the part of the male. But it will be forced into the female role if it meets up with an individual with even fewer dots on its plasma. . . .
>
> In the same way, a rather weakly developed female can act as a male when faced by a strongly developed animal. It can even happen that three individuals copulate with one another, the middle one penetrating the first as a male, while at the same time serving as female for the third.

Once the two "genders" have come into contact the male gradually works his way into the female's body (or is drawn in, like food) until, bent and cramped, he is completely inside.

Then he simply dissolves (or is eaten). His cytoplasm merges with that of the female. All that is left of the male is his genes, which are drawn upward through the female's body to her "head," where her own genes are stored. There the recombination and gene-shuffling characteristic of sex takes place. Four new gene sets are made, each one of which is drawn off into a different quarter of the super-*Trichonympha* that was made up from the male and female combining together. Then it (they? she?) divides into four small creatures, thus providing each set of genes with a cytoplasmic envelope.

This all takes about ten to eleven hours. Meanwhile the cockroach has proceeded with its molting. The new generation of sexually produced *Trichonymphae* are expelled from their host and rest dormant in the excreted lining. There they will stay, eventually dying unless another cockroach happens to eat the molt of the first. If that happens, then the sexual progeny will start to reproduce asexually until the next eviction notice is posted.

A slightly different example of sex among the protista is that of a small alga, about the size of a red blood cell, called *Chlamydomonas reinhardi* (clam-a-da-MON-as). This is one of those creatures that made problems for the taxonomists working under the old two-kingdom, plant-or-animal classification, since it both carries chlorophyll and is mobile, driving itself toward light sources with two "arms": strong, long flagella that pull it forward breast-stroke style. All it needs to support itself are sunlight, water, some dissolved salts, and a steady supply of nitrogen which it gets from ammonia in the water. When those resources are all available it reproduces asexually, splitting from top to bottom every eight to ten hours.

The algae only switch to sex when the supply of ammonia is exhausted, or withdrawn by a meddlesome experimenter. There are two genders (though microbiologists do not call them male and female, but, somewhat more neutrally, "+" and "–"). All matings involve one of each. When two receptive algae pass each other in the water, they bring their flagella into contact. If the two algae are of the right sort, the tips fuse and yank the two organisms together. Under a microscope the algae seem to swim at a terrific velocity, and this fusing-and-yanking process, which appears to happen instantly, can look very violent. Once they have been jerked together they both release an enzyme that cuts away their body walls. The "+" cell extends a small projectile tube that connects the two and forms a cytoplasmic bridge. The bridge widens steadily, pulling the two cells together, and a single "double" cell results. The flagella then unstick and the four-tailed double organism swims off, probably searching for a good place to "hibernate." After a period it (they) forms a hard spherical coating around itself (themselves) and awaits the return of favorable circumstances—usually the right level of ammonia in the environment. *C. reinhardi* have lasted in this condition for at least four years. When it does detect the right conditions the alga shuffles its genes and divides itself into four progeny among whom all the cellular proteins are distributed. The spore hatches and the four small algae swim away. The parents are not quite dead, since in a sense they have transformed themselves into their four progeny; and they are not quite still alive, since the sexual process has made their progeny genetically distinct from their parents.

These two examples of protist reproduction show that the switch to sex is often associated with the destruction or collapse of the habitat in which the asexual reproduction went on. Microbiologists find that often they can induce sex at will in laboratory cultures by letting, or making, conditions deteriorate: by lowering the culture's temperature, or letting it dry out, or by withdrawing vitamins or some other nutrient. On the other hand sex can be forestalled, often seemingly forever, by tenderly ministering to the organism's needs. The changes that induce sex would, in the real world, be events like the coming of winter, the evaporation of a pond or puddle, the exhaustion of a nonrenewable nutrient—changes that would be catastrophic and irreversible. From a distance protist sex often looks literally like a last fling, something engaged in by the last generation to live in a habitat that has supported that population for many generations.

That point is relevant to the inquiry at hand, but it is really not what is most interesting about protist reproduction, which is the overwhelming variety of it all. Two examples do not begin to do justice to their zany imagination. Many reproduce by mutual immersion, like *Trichonympha*; others, less headlong, form a cytoplasmic bridge between the two individuals, over which the gene copies migrate. Usually this exchange is reciprocal, with two gene copies passing each other over the bridge, but sometimes it is just one-way. Then the participants separate and continue an independent life. Sometimes they incorporate the new genes into their own functioning set, the way some bacteria do, and in other cases they hold the new set in storage until the time comes to make offspring. Sometimes two *Paramecia* will undergo all the preparations for mating—meeting, aligning, joining their surfaces, building a bridge, making a copy of their genes—only to stop just short of the actual migration itself. They then destroy the original gene sets and recreate their nuclei from the copies. Thus the interaction, ostensibly for mating purposes, was really a way of mutually stimulating each other into a genetic housecleaning of some kind.

Asexual and sexual reproduction can be combined in many ways. There is one protozoan (*Polyspira*) that lives in the molted skin of hermit crabs. Individuals first join into pairs (called *syzygies*) which look vaguely like yin-yang symbols. Then they split, asexually, again and again, each time reducing their volume by half, until they end as a chain of six or eight pairs of much smaller yin-yang symbols, arranged like beads on a string. Only when this division has stopped does the sexual exchange take place between members of the small pairs. Then all the individuals split up and disperse. Some biologists call this *syndesmogamy*; others call it *zygo-*

palintomy. These terms at least show what a supermarket of prefixes, roots, and suffixes are needed to label the strange varieties of protist reproduction.

Continuing to move along the sexual spectrum brings us over another dividing line, that separating the single-celled from the multicelled creatures. Many of the larger multicelled organisms are committed to sexuality. There are very few asexual amphibians, reptiles, or fish; no known asexual mollusks, which is a huge group of species comprising all shellfish, snails, squid, and octopuses. There are no asexual birds or asexual mammals (the instance recorded in the New Testament is thought to be unique).

But some of the smaller multicellulars still retain the asexual option; one of these is a powerful little protist predator called a *rotifer*. While these animals are still too small to be seen easily, they represent another big "advance" in complexity over the protista. In general they are squat, chunky creatures that look a little—a very little—like those small, tubular, electric coffee grinders, but with their blades exposed.

Most rotifers have hard, pointed jaws that can sometimes be extended to grasp and chop into their prey. They often have body armor in a wide variety of shapes and types. They have a brain, a complicated chewing apparatus, muscles, digestive glands, a nervous system, a urinary bladder, and, sometimes, a foot with two toes that they use to help themselves move over certain surfaces, humping along inchwormlike. Some rotifers are even capable of major bodily alterations. There is one species (a *Brachionis*) that is normally spineless. In this vulnerable condition it can find itself being hunted by a second species of rotifer. If the predator rotifer is introduced into water samples containing the prey, the next generation of the *Brachionis* to appear will have spines; the length of the spines will depend in part on the number of predators in the parents' environment.

Like the single-celled protists, rotifers seem to prefer to reproduce asexually. Females lay unfertilized eggs that hatch into more females. In some cases, however, some species of rotifers can be induced to make males and reproduce sexually. These males are usually dwarfed and highly simplified (their digestive systems never develop), and have life-spans of not more than a few hours. They are adapted to find and fertilize a female—quickly; nothing more. A female making these little males will continue doing so until she is mated herself (not necessarily by one of her own progeny, of course). She will then switch back into making females again; the only difference is that her next clutch of daughters will hatch from fertilized eggs.

One way of inducing sex in rotifers is by letting the environment deteriorate (by starving them, for instance); this is the same "last-fling" effect noticed with the protists. John Gilbert, a zoologist at Dartmouth, has found that rotifers will also switch to sex when they are crowded into dense populations. Crowding them with protists does not have the same effect; it has to be with numbers of their own kind. Gilbert points out that male rotifers are more often found in nature at high population densities as well. He believes that they census their populations by monitoring the concentration of a chemical they release into the water.[20]

Other multicellulars show these same two associations with the switch to sex (crowding and habitat collapse). The gall midges are tiny flies that feed on mushrooms and other fungal growths. When this fly finds a mushroom—such a food source might be hundreds of thousands of times its size—it will multiply, asexually, at a breakneck pace. It does this by drastically reducing the time needed to reach a reproductive state. The young of each succeeding generation grow inside the mother, so that, as she eats on the mushroom, her young eat her up from inside. (A more accurate way of putting it might be to say that she transforms herself into her young from the inside out.) By the time the larvae emerge, the mother has been reduced to a hollow shell, and within two days, the next generation will be consuming the larvae in turn. Gall-midge populations multiplying like this have reached 20,000 per square foot in five weeks.

That is the asexual phase.* There is also a sexual form, which is winged and egg-laying (the young develop outside the mother). And just as with the rotifers, there are two kinds of effects that trigger the development of this sexual form: starvation and crowding.

As one continues to move along the sexual spectrum, two observations stand out. The more sexual a creature is the larger it is (and the more long-lived and physiologically complicated), and the more likely it is to live in an environment that is crowded with members of the same species over the long run. Asexual marine organisms are more likely to be found in the ocean depths than in the biotically richer, more crowded, surface. (Similarly, there is a European flatworm [*Planaria alpina*] that is sexual when it lives in rich, warm climates but reproduces asexually in Scandinavia.) Asexual sea anemones tend to live in shallow water, where temperature, salinity, and degree of exposure to the sun's rays all vary greatly; these pools may get crowded from time to time, but only over the short run. The sexual anemones live lower down in the intertidal zone, where the pools are deeper and more stable conditions promote a more extended competition. The few known asexual vertebrates seem to be almost all animals that specialize in temporary habitats created by fire, flood, drought, or tree blowdowns. Most of the asexual lizards, for example, are found either along the edges of

* This principle of beginning to reproduce while still a juvenile and doing so by converting one's internal tissues into offspring, is also followed by the best-known asexual insect, the aphid. In this case embryonic development begins in aphid mothers before they themselves have even been born, so that if one looks at the right time one can see two generations telescoped within a single aphid grandmother-to-be. It has been calculated that in one of these species a single individual can transform herself into 524 billion aphids in a single year. Aphids also have a sexual, egg-laying form which is triggered, at least in those species that live in temperate latitudes, by a decrease in the length of the day. They then mate and lay eggs which overwinter until the spring.

forests, or where trees have been disturbed or destroyed, or in the floodplains of rivers, streams, and washes. A number of these species are known in the American Southwest, where the rough terrain, sparse vegetation, and extremely uneven precipitation make flash floods particularly destructive—especially to something as small and vulnerable as a lizard. It has been speculated that construction in the Rio Grande Basin has, since 1941, so effectively controlled flooding that a number of asexual populations have been wiped out.

As was mentioned briefly in the last chapter, the theory of sex these pages rely upon is that sex is selected when and as it becomes important for an individual to have a diversified set of offspring. An asexual organism can only reproduce itself; if it has specializations, all its progeny will have to pursue the same trade, practise the same tricks. But a sexual creature, by joining its genes with another organism (and at the next mating go-around with a third, and then a fourth), expands the range of vocational orientations that are passed on to its offspring. To put it another way, sex ought to be selected as it becomes more important to a parent to have offspring that are distinct individuals, different from their siblings and their parents.

If this theory is right, then questions about the role of sex transform into questions about the importance of specialization and individuation. Judging from the evidence in hand so far, across all species as a whole, there should be some reason why individualism becomes more important as species get larger, live longer, and as they live in more stable and crowded environments; while, within just those species that can switch modes freely, having varied progeny would become important when a habitat is about to collapse, or when it gets crowded.

In trying to spell out why these links should exist I have found it helpful to imagine a second way of comparing all species against each other (in addition to the sexual spectrum). This might be called the means-of-population-control spectrum. All species reproduce to excess, way past the carrying capacity of their niche. In her lifetime a lioness might leave 20 cubs; a pigeon, 150 chicks; a mouse, 1,000 kits; a trout, 20,000 fry, a tuna or a cod, a million fry or more; an elm tree, several million seeds; and an oyster, perhaps a hundred million spat. If one assumes that the population of each of these species is, from generation to generation, roughly equal, then on the average only one offspring will survive to replace each parent. All the other thousands and millions will die, one way or another. This "excess" fecundity can be pruned back in only two ways. One is by competitive interactions with other members of the same species (by which I do not mean fighting so much as another member's getting to a needed resource first, though of course direct, physical competition counts too). The second way is through fluctuations in the ecology, the depredations of predators and parasites, and, in general, the influence of natural forces or creatures that are very different from the affected organism. The first type of species controls its own population growth; the second depends on nature to do it. To put it another way, all living things are victimized either by nature or by society, and they are all victimized to the same degree, until their populations have been cut back to replacement level.

Most biologists believe that the larger an animal is the more important social competition is likely to be in that species' scheme of population control. One way of arguing this is quite indirect; it says that larger animals are inherently better at coping with ecological fluctuations, because if an unpleasant situation comes up they can physically remove themselves to search for a more pleasant one. They have more powerful swimming surfaces (or legs), more bodily mass with which to retain heat and conserve momentum, and more internal volume for reserves and stores and specialized organs. Since excess population has to be trimmed back either by nature or by society, therefore large creatures must be relatively more vulnerable to social competition. Or one can argue that large creatures are more likely to run across each other and influence each other's fortunes. In fact size itself is thought sometimes to be an adaptation to social competition, since a good big guy, as the proverb goes, will always beat a good little guy.

To return to a point made earlier, the world that small creatures, those with boom-and-bust population cycles, live in is a uniform one. It has only two states: abounding plenty and unrelieved famine, and the latter replaces the former in a blink. The reason is that in populations with soaring growth rates every generation but the very last one, the one for which everything comes to an end, enjoys abundance. If the population doubles with each generation, the parents of the last one of all will, as they start to breed, gaze out contentedly on a world that still holds as yet unexploited a full 50 percent of all the resources it has ever contained. (This point is sometimes made to those who believe that the existence of large oil reserves undercuts the claim that there is an energy crisis.) Ecologically organisms on the bacterial end of the scale live in a world that is almost always lush and ample and rich. When the niche closes up—perhaps a bigger competitor has arrived, or the population/resources ratio becomes impossibly unbalanced, or the accumulation of wastes reaches toxic levels—it does so quickly, for everyone at the same time. The organisms all flip into dormancy, into seeds and spores and other low-energy states. (It is while they are seeds or spores, for the most part, that their numbers are cut back; more often than not, the favorable environment they need does not return until even the seed's formidable powers of endurance have been exhausted.) Finally, their world is socially very uniform as well. They are surrounded by members of the same clone, asexually produced, genetically identical siblings, and all these siblings are doing just two things—gobbling and splitting, gobbling and splitting, over and over, as fast as possible.

By comparison, large creatures, which live in, forage over, and grow through a succession of diverse environments, live in a world with a rich texture of problems and possibilities. At the same time they are being pressed by the forces of social competition, which refers to the problems that a population of interacting organisms, all with the same needs and after the same resources, can make for each other. (Bacteria-type organisms do not have these problems, by contrast, because by the time resources get short enough to be fought over the whole habitat has become unendurable.) What kind of adaptation will be selected by socially competitive creatures living in a multitextured world?

The answer is obvious: individual specialization. One sees the same result looking at human society. Specialization has gone much further in stable, crowded, enduring, competitive societies (like New York!) than in either boom-and-bust populations (like the crowd in a stadium) or in less crowded, rural, areas. Theoretically specializations in nature can involve a concentration on any feature of the life-style: being specially skilled at finding one kind of item in a species' diet range rather than another, or being extracompetent at converting certain foods into cellular materials. One can imagine specializing in certain specific foraging styles. Some members of a species might be good at being bold and aggressively pursuing suggestive possibilities, while others might be good at being cautious, and steady, and sticking to the well known. Another fraction of the population might be professionals at stubbornly coping with a slowly deteriorating environment, while others may be expert at how best to search for a new one. Some may forage best when and where the food is rich but the number of competitors high; others where the country is poorer but the social circumstances are more relaxed.

So the theory is that specializations are defined and pursued when a socially competitive species lives in a world with some variety to it. Is there any direct evidence for this theory? Some, but, unfortunately, not much. It is hard to get firsthand observational evidence on this subject—to figure out what the professional structure of a species might be—because all the members of a species (usually) look the same, regardless of what specialization they might or might not be following. They have no distinctive occupational titles or cues. If they did it would be possible to associate differences in the way two animals live with differences in their labels and maybe draw some conclusions from that. Evidence like this is much easier to get by looking at what happens among individuals of different species, because they do look different. Here we do see this principle—that competition forces specialization—confirmed; in fact, the confirmation is so powerful that it is practically a law, and is stated thus: No two species can exist at the same locality if they have identical ecological requirements. This is called *the exclusion principle*, and it is based on the belief that if two species try to live in the same way at the same place, one, by virtue of slightly superior specialization to the conditions at hand, will always exclude the other.

There are many famous examples of this principle; one of the best-known is that of the fauna that live on cow dung. One study showed that of fifty-one different species of nematodes (worms) that were found to eat bacteria and fungi in cow dung, none seemed to be in direct, head-to-head competition. All had their own specialization, their own trade. Some lived on the surface, others at different depths of the interior, or at different stages of decomposition, or different levels of moisture, and so on. Arguably if any one of those fifty-one species were to find itself in sole possession of a piece of cow dung, then it could manage, however clumsily, to forage over a much wider range of conditions than it does now. But the competition of the other specialists forces it to stick to what it does best. A jack-of-all-trades-but-master-of-none nematode that wandered vaguely about trying one thing here and another there, could only sustain so inefficient a foraging style if its cow dung was lush and rank and overgrown. In real life the average piece of cow dung is harvested much too meticulously by the specialists for it ever to lapse into so abandoned a condition. The theory being used here says that competition within a species has the same effect. Specialists emerge, divvy up the territory, and drive out the generalists.

Nonetheless, for all the difficulties, there is some direct evidence on intraspecific specialization. There is a common experiment with fruit flies in which different varieties of the same species are placed together in a cage with a mix of foods.

The flies are then allowed to breed for a number of generations; with each generation the proportion of the whole population made up by each variety is recorded. Usually these proportions vary dramatically: Variety A might be dominant for a while, then B, and then A again, and then C, and so on. These fluctuations are believed to reflect the changing proportions of the different foods on which each variety has specialized. Variety A will be dominant for a few generations because its preferred food happened to be most abundant when the experiment began, but, as its numbers multiply and the amount of that kind of food declines, a point is reached when the food on which B specializes becomes more common and B becomes dominant.

These specializations need not always have to do with foraging. Pine trees can be infested by little parasites called scale insects. Botanists have reported that the intensity of infestation by these insects varies enormously from one pine to another. One report notes:

> Scale-free pines frequently stand for years beside trees infested with as many as ten insects per centimeter of needle, often with intertwining branches. Scale-free trees tend to remain uninfested even though . . . larvae can be seen crawling on their needles during the insect's reproductive period in July. When plots of trees are sprayed with insecticide to control the scales, trees are reinfested . . . to approximately their original level of infestation; that is, formerly severely reinfested trees become severely reinfested, previously lightly infested trees become lightly infested . . .[55]

Apparently individual trees differ in which kind of anti-insect toxin they produce, while the insects differ among themselves as to which toxin they can neutralize. The trees have not been able to evolve a toxin that will protect against every scale; the scales have been unable to devise a neutralizer that will render all the toxin varieties harmless. Instead of either one of these solutions we have two armies of specialists parrying with each other.

Occasionally one finds species in which the specialists are

visibly different, though such finds are rare. Recently a group of Mexican fish (*Cichlasoma*) that were thought to be comprised of several different species were reexamined and identified as a single species. These varieties differ in tooth structure, body shape, gut length, and diet. One variety eats snails and has teeth adapted for crushing shells; a second eats algae; and a third eats other fish. The single-species identification was made on the basis of genetic analysis and the observation that these three varieties were found being raised in a common brood.[125]

It cannot be emphasized strongly enough that all the issues of specialization are social issues. The argument has already been made that specializations themselves only evolve when there exists a society of interacting competitors. But even when specializations have defined themselves, whether or not they work for the organism will depend solely on what everyone else in the species is doing. Specifically, the more individuals there are in a given vocation the worse life will be for those specializations. The more flies there are in Variety A, comparatively, the rosier prospects will be for the Variety B flies, since they will have fewer competitors in their specialization. The fewer neighbors that a pine tree has that mount the same anti-insect defense that it does, the better; otherwise, if one of those neighbors' defenses is penetrated by an insect, then that pine tree will be caught as well (by that insect's progeny). It is far better for a tree to be the only specialist of its kind in a grove. (The same point can be made about gender specializations. The fewer males there are, the better it is to be one, and the more, the worse.)

In other words being a specialist is not necessarily a good thing; if an organism has too many fellow-practitioners in its specialty it might end up even worse off than if it had been a generalist. At least generalists have some flexibility; a specialist in a crowded profession is condemned to a frantically competitive life by the very rigidity of its specialization.

Faced with all these considerations, what's a parent to do? She can't make a single, all-purpose model progeny, because offspring like that would be outcompeted at everything they tried. Nor is turning out just one kind of specialist a good idea. In the first place, all her offspring will be competing against each other, but, more importantly, the parent would be putting all her eggs in one basket. Who can tell where the opportunities are going to lie by the time her progeny have grown up? Far better to cover her bets, produce a mix of specialists, and so position herself to profit from opportunities arising anywhere on the vocational spectrum. And the technique that turns that trick is sexual reproduction.

Sex therefore is a social adaptation, springing from and responding to social dynamics at every point. It follows naturally why it should be associated with stable, crowded populations. The correlation with habitats poised on the brink of dissolution is less obvious. It may be that these small, mostly one-celled parents compete over who is going to get the first jump over the others when the next good environment appears. The first algae, say, to begin reproducing asexually in a new habitat has a good chance of multiplying so fast that it can take over much of the puddle all by itself, so that the habitat is filled just with descendants of one or two designs.

An alga spore with no competition would take its own sweet time about waking up and beginning to swim and split; it would wait until it was absolutely sure that good conditions had in fact appeared. But a spore with competition does not have this luxury; it must guess and run the risk of opening prematurely. A parent might then be expected to build a range of progeny, some impetuous and daring, others temperate and judicious, still others very conservative and extremely slow to test the waters. The only reason it diversifies its offspring like this—an act which insures that most of its progeny are going to be inappropriate for whatever environment does appear—is because all the other local parents are doing the same thing. When these smaller organisms reproduce sexually, they are positioning themselves to deal with one of the few moments in their lives when the society of others of their kind is important to them, and that is the scramble to be first over the threshold into a new environment. The rest of the time they live asocially, and therefore asexually.

So, the theory of sex is that social competition selects a society of specialists, and specializations themselves select sex, since it is advantageous for a parent to breed a little society of specialists herself. The critical idea of this chapter, the one that has the most importance for the chapters to follow, is the vision of sexual societies being composed of interacting guilds of professionals. And at the core of that idea should be the image of each member of these civilizations being highly specialized, individualized, and competing by being good at what he or she does—by harvesting efficiently and skillfully—rather than through physical confrontation.

CHAPTER

VI

THE CONSEQUENCES OF BEING SEXUAL

IN THE LAST CENTURY A FAMILY OF MARINE JELLYFISH called the Siphonophora (the man-of-war is one) became involved in nineteenth-century socialist thought. Naturalists studying these animals showed that each jellyfish was really a tightly coordinated ensemble of several separate individuals, all working toward the common good. Every one of these "component" individuals had its own nervous system, digested its own food (which was delivered to it through a communal pipeline), and often had other features of self-sufficiency, depending on the species. But each "individual" also worked cooperatively with the others. For example, one might form a balloon that kept the whole colony afloat and drive it before the wind; a second might pump water to propel the colony; a third might prepare the material to be delivered down the communal pipeline; a fourth might form tentacles and look out for the defense of the colony and the capture of prey; and a fifth might occupy itself with making the colony's next generation. The socialist community seized on this phenomenon as a parable of the utopian state. "As in a communist state," one lecture ran, "there are here no poor by the side of the rich, no hunger beside surfeit, . . . no lazy next to the industrious. Each one contributes his part to the existence and welfare of the whole. . . ."

What the Siphonophora teach us is that there is more than one way to be a specialist; it is perfectly possible to be a specialized member of a large organization. But sexual creatures tend not to be this kind of specialist; they are generally free-living, independent, self-sufficient individuals. "Sex is an antisocial force in evolution," E. O. Wilson says flatly in his magisterial book, *Sociobiology*, by which he means that tightly integrated organizations tend not to be made up of sexual organisms. The Siphonophora have a sexual stage, or state, but it is not a colonial individual. When sex cells made by two Siphonophores of the right species join, they form a solitary, single-celled "seeker" that swims about looking for the right time and place to begin a colony. Then it buds off asexually all the cells that make up the colony, so that each colonial cell is genetically identical.

A second test of organized solidarity (besides a collective of specialists) is altruistic sacrifice—the laying down of one's life so that others might live. A dramatic example of this—one that would delight the German socialists as much as the story of the Siphonophora—is that of a social fungi (*Dictyostelium discoideum*) called the slime mold. Part of its life it lives as a free-living, independent, amoeboid individual, roaming through the debris of hardwood forests, foraging on the bacteria that in turn feed on decaying wood. These cells split, reproducing asexually, every three or four hours, multiplying at this brisk pace until the population of bacteria has sunk to very low levels. At this point one or two of the amoebas in the area begin making and transmitting pulses of a certain hormone. This hormone (sometimes called *the alarmone* because it is involved in starvation reactions in a great many species, including humans) spreads out in circular ripples. When it hits other amoebas something about the angle of the ripple-body contact tells the receiving amoebas where the sender is, and they begin to crawl in that direction. At the same time each receiver puts out a pulse of alarmone itself; this pulse is timed such that it will fall into synchrony with the first ripple. Small amounts of a chemical that destroys alarmone are also released; this prevents the "noise" of backwash eddies of the hormone from distracting the congregating amoebas. Only a direct pulse is large enough to override the effects of this "muffling" enzyme.

As the ripples of alarmone spread out through the neighborhood hundreds, and then thousands, and finally tens of thousands of amoebas converge on the first sender. They line up in circling streams that spiral in on the point of aggregation like the arms of a whirlpool; each time a ripple sweeps outward all the amoebas affected by that wave take one "step" inward, amplify and relay the ripple outward, and then rest. Moving amoebas appear brighter than resting ones, so that under a microscope and with the right kind of illumination one sees concentric, alternating bands of light and shadow encircling the center. As time goes on—this whole process takes eight to ten hours—the pulses of alarmone come faster and faster. The dark bands, indicating resting amoebas, shrink steadily, and the whole joustling throng seems to be in continuous motion. As the amoebas aggregate they pile up in the center, forming a tiny, upright, fingerlike body. Somehow the colony as a whole senses how big it is getting, because when it has recruited about one hundred thousand cells, and grown to about one-twenty-fifth of an inch long, the finger topples over on its side. The slug orients on a source of heat and light and begins to crawl toward it. It does this by comparing the amount of energy received on its front end with the amount received on its back end and directing itself so that the front end always receives more. (The slug can perform these calculations with such uncanny accuracy that it will orient and crawl toward a spot of fluorescent paint dabbed on a laboratory wall.) In nature this behavior will bring it out from under dirt and leaves into the open. A slug is capable of pursuing an energy source, at least in the lab, for as long as ten days.

When the colony senses that its energy source is overhead, it reshapes itself into a Mexican-hatlike form. Eighty percent of the slug's component cells change into hard, dry, resistant spores. The remaining 20 percent elongate themselves vertically, boosting the spores up into the air, as though a glassy lemon were being lifted by a rapidly extending fiber-glass rod. These stalk cells form a tough, cellulose scaffolding around their outsides and then die. The spores contained in the head wait to be swept off by wind or water to more fruitful conditions. The twenty thousand cells that made up the stalk have died so that the other eighty thousand can have another turn in their life cycle. What better example could one have of a noble, self-sacrificing, generous devotion to the common good?

Stories like this involving sexual creatures are hard to come by. One can't say flatly that sexual organisms never sacrifice themselves to the interests of others, and never form organizations of mutually dependent specialists. There are exceptions, but in most cases even these seem to have made a special effort to sidestep the influence of sex. One such set of exceptions are the social insects—the social bees, ants, and termites. Colonies of these insects usually are composed of two or three different castes, or specialists; the altruism of these workers, their willingness to fling away their lives in defense of the colony, is famous; and such nests typically depend on the efforts of castes of workers that, while technically female, never lay eggs of their own. (From a Darwinian point of view this is just as self-sacrificing as a bee worker plunging her stinger irretrievably into someone she sees as a threat.)

These insect colonies are wonderful entities, but one can exaggerate how organized they are. E. O. Wilson writes:

> An important first rule concerning mass action is that it usually results from the conflicting actions of many workers. The individual workers pay only limited attention to the behavior of nest-mates near them, and they are largely unaware of the moment-by-moment condition of the colony as a whole. Anyone who has watched an ant colony emigrating from one nest site to another has seen this principle vividly illustrated. As workers stream outward carrying eggs, larvae, and pupae . . . other

workers are busy carrying them back again. Still other workers run back and forth carrying nothing. . .

Obviously the image of an anthill as a tightly regimented society in which, to use T. H. White's phrase, all that is not mandatory is forbidden, is exaggerated. Much can be accomplished by these inefficient, freewheeling, quasi-democracies, of course, but they are simply not in the same league for disciplined and elaborate organization that the asexual cells of our bodies are. We have dozens of specialist castes (which in our case we call specialized tissues), not just two or three.

But even as comparatively simple an organization as a colony of social insects has found it necessary to suppress sex. Eleven of the twelve families of social insects make their males asexually. When the queen lays an egg she can choose whether or not to withdraw a sperm from the sperm bank she received when she was fertilized. If she does fertilize the egg with a sperm, then the egg becomes a female; if not, a male. Males, in other words, have no fathers, only mothers; females have both.

Because the males have no fathers they inherit only one set of genes, their mother's. And when they get around to making their sperm they have just this one set to draw from. With only one set of genes there can be no variation among their sperm—all the sex cells any individual male makes during his short life will be identical. (By contrast, a sexually produced male, like a human, would build his sperm by mixing together his mother's and his father's genes; almost certainly each time he does this a qualitatively different sperm will result.) When that male mates with a queen-to-be, he leaves her with a large number of identical sperm, all of which she stores in her sperm bank. The queen uses these sperm to make daughters. The result is that the females in most social insect colonies resemble each other much more closely than sexually produced sisters usually do; as far as the genes they got from their father are concerned, they are all identical twins.

The reason that social insects (eleven out of twelve times)

have found it necessary to suppress sex is probably that individuality is a problem for any organization. Members of an organization are interdependent; there is a general reliance on all members routinely discharging their particular duties. Since sexuality seems intimately involved with the thorough distribution of individual differences in a population, it is easy to see why it might be suppressed or never allowed to appear in an organization.

Sexual creatures do often pool their efforts by flocking, migrating, and/or breeding together; combining their powers of observation and/or resistance makes them all safer. A nineteenth-century biologist, Francis Galton, put the matter thus:

> To live gregariously is to become a fibre in a vast sentient web overspreading many acres; it is to become the possessor of faculties always awake, of eyes that see in all directions, of ears and nostrils that explore a broad belt of air; it is to become the occupier of every bit of vantage ground whence the approach of a lurking enemy might be overlooked. . . .

It has been suggested that little fish combine into a school so as to disguise themselves as a single giant fish, large enough to be their predator's predator, and so scare him away. A second explanation of the point of cooperative assemblies is that when prey animals cluster they become harder to find, because they are dispersed less evenly in space and time. Predators have to spend more time searching or waiting for them; this inconvenience reduces the number of predators and therefore, the number of attacks that the flocking animals must suffer. Galton's point—that flocking animals profit from combining their powers of observation—is still widely cited. Flocking also seems to be the best strategy to pursue when harvesting a wide range of foods.

The point to be made here is that sexual members of these associations keep strict limits on the degree of dependency they allow to spring up between them and the rest of the flock. Symbioses among sexual members of the same species do not even come close to the tight, intimate, mutual dependencies that we see in such classic symbioses as that of the fungi and algae to make lichens, or that of the clovers and bacteria to fix nitrogen (make natural nitrogenous fertilizers). The institution of the community specialist that we saw so well exemplified in the jellyfish Siphonophora, wherein one member concentrates all its energies on one phase of the life cycle while being completely dependent on others to support it through the other phases, is virtually absent among sexual organisms. By and large associations among sexual organisms form when it becomes profitable to pursue, as a flock or herd or group, the same sorts of activities that each one would be doing if it were alone.

Typically the members of a group of sexual animals compete among each other on some issues while they cooperate on others. A good example of this was found recently in a study of communally nesting birds, in which more than one mated pair join forces in a single large nest. These pairs cooperate to guard their young jointly; they brood their eggs and feed the fledglings with no obvious bias toward their own offspring. Darwin, and others since, have noticed that such nests often have eggs strewn about them. Explanations for this waste usually invoke a presumed failure of some kind in the breeding system; Darwin himself attributed it to male incompetence at brooding. Recently Sandra Vehrencamp of the University of California investigated this egg waste in one such species, the groove-billed ani (*Crotophaga sulcirostris*), in Costa Rica.

Vehrencamp found that the females were rolling each other's eggs out of the nest! She believes that there is a limit to the number of nestlings that any group can brood, feed, and defend, and that females compete within the cooperative framework of the communal nest to see who can acquire the largest share of these "nestling slots". A female will only roll eggs out of the nest until she begins laying herself; apparently (and incredibly) a female cannot tell her own eggs from those laid by other females, so to avoid the risk of tossing out her own by mistake she then stops rolling them out altogether. Therefore, early-laying females lose more eggs than late-laying females. The female to lay last loses none, and stands to garner the largest proportion of nestling slots for her own offspring. The early females reduce this benefit to the last layer in two ways: (1) They lay more eggs, including one late in their cycle, frequently at the same time the last female begins to lay, and (2) they also lay more slowly, spreading their eggs out over longer periods. Vehrencamp believes that ani society is structured by an age-dependent, female-oriented status hierarchy in which the oldest females get both the largest males—the ani is monogamous, like many birds—and the most fledgling slots. The males' incubation effort is a function of the social position of their female; the higher she is the harder they work—and the less work she does, at least as defined by rates of brooding and feeding the young.

Feeding a flock of pigeons gives a glimpse of the same phenomenon: a high level of squabbling competitiveness within a framework of cooperation on other issues (one issue on which flock members cooperate is a common flight path).

The other test of nonindividuality that has been mentioned here is altruism, the sacrifice of one's own interests to support those of another. As anyone who has pondered the question of genuine unselfishness knows, it is sometimes hard to tell the ethical difference between sacrificing for oneself (selfishness), and sacrificing for others with whom one has everything in common. Is that genuine altruism, or is it just another form of selfishness? Biologists have not really resolved this problem in their own sphere either. Take as an example the relation between mother and child. Mothers generally are willing to sacrifice a great deal of their time and energy for their offspring. No one calls this altruism, since it is sacrifice in the cause of reproduction. The mother is working to perpetuate herself in the next generation; her interests, and her progeny's, are very close to being identical.

On the other hand the word altruism is used in describing the sacrifice of the slime-mold cells that transformed themselves into the stalk. Yet those cells' sacrifice is really no different than that of mothers; since the slime-mold amoebas reproduce asexually, all the organisms likely to congregate into a single slug are identical to each other. In a sense, therefore, it is no sacrifice at all for those twenty thousand cells to die helping the others, because the ones that die *are* the others.

The point can be reasoned out even more rigorously. How can an altruistic trait spread through a population? Why doesn't self-sacrifice work against itself in evolutionary competition? One answer is to imagine that there are two copies of the altruistic genes in the population, and that the individual with one copy devotes his altruism to the individual with the other. When such a partnership makes sense, "unselfishness" will then defeat "selfishness" in the Darwinian race. A mother has, comparatively, an aging body that is going to die soon anyway; it is far better for her genes to help their copies in the fresher, younger body of her offspring than to try to help themselves to eke out a few more weeks or months. When the slime-mold habitat collapses, all the amoebas will die if they stay where they are. If only a fraction can escape, and forming a partnership allows more cells to escape than otherwise could, then genes predisposed to the altruistic behavior will outcompete more individualistic ones. All that is necessary for this dynamic to work is that the donor and the beneficiary of the aid have much in common genetically.

Sexual reproduction, because it mixes the genes of two parents, decreases the prospects of that mutuality. All other things being equal, it inhibits the development of self-sacrifice and promotes the evolution of self-assertiveness and competitiveness. Imagine a gene appearing in an asexual organism that promotes a willingness to share (or a less intense grabbiness). When that asexual organism clones itself, all its offspring will carry that mutation, and there will be a certain chance that one individual holding the gene coding for that behavior will encounter and therefore help a sibling that also has a copy of that gene. But since a sexual organism has had to mix its genes with those of a stranger to make offspring, and since the genes of each parent are equally represented in the next generation, the chances of an altruistic gene being able to help itself in a meeting with a sibling are cut in half.

The point can be restated by looking at aggressive behavior. "Selfish" behavior that arises in an asexual family has a certain chance of hurting itself and hindering its spread because it hurts other members of the family that carry the same gene. But under the same circumstances, the chance of aggressiveness getting in its own way is halved in a system of sexual reproduction. The conclusion is that the evolution of altruistic behavior is half as likely, and the evolution of socially aggressive, individualistic, behavior twice as likely, in sexually reproducing creatures.

Obviously all this can be taken too far; altruism does exist among sexual animals. But many biologists believe that such exceptions only will be found among very close relatives—parents and their offspring, brothers and sisters, and much less commonly, among uncles and aunts and nieces and nephews—and that even in very closely related creatures some degree of competitiveness is likely to be found. This whole theory is called the theory of *kin selection*, and it has proved very useful in explaining patterns of animal behavior. Robert Trivers and Hope Hare of Harvard have used kin selection to explain the marginal position of males in social insect societies. For centuries it has been known that all the business of these colonies is kept tightly in the hands of females. They care for the eggs and nourish the developing young, construct the nest, keep it clean and well ventilated, guard it against enemies, and repair it when the nest is damaged. They scout for food, fight for it if need be when it is found, and bring it back to the nest. They mark off, defend, and maintain their nests' territories. If they are ants that garden, it is the females that tend the crops; the female army ants pillage and sack everything in the path of their marauding columns; female domesticators tend their aphid herds. Males never take part in these activities; something in the evolution of the sisterhood has excluded them. They are made in small numbers, cannibalized in times of protein shortage, and driven from the nest by force when the sisters decide the time has arrived for their mating flight. (Apparently the males' restricted role has left them so stupid that they don't even know the best time to look for females.) An early entomologist, Forel, remarked with some vehemence ". . . The males (of ants) are incredibly stupid . . . the organs of thought are very large in the workers, much smaller in the queen, and almost wholly atrophied in the males." Trivers and Hare measured the amount of food fed by female workers to their reproductive sisters (the insects that are going to fly off, mate, and try to found new colonies) compared to that given their reproductive brothers. Their finding was that the females fed their sisters three times more than their brothers. Since females are three times more closely related to each other than they are to their brothers, their result shows that altruism (in the social insects) exactly parallels degrees of kinship.*

Kinship theory has also been used to explain a phenomenon called the helpers at the nest. These are animals, found in a wide variety of species, that seem to devote their energies to helping others of their species breed. An ornithologist named Woolfenden marked the helpers found in the Florida

* Each sister has two sets of genes, one from her mother and one from her father. As explained earlier, it is likely that all the sisters in the nest have inherited exactly the same set of genes from their father, so 50 percent of the genes of every sister is likely to be given over to a single, uniform copy of the paternal inheritance. Since mothers develop from fertilized eggs, unlike males, they have two gene sets themselves, and in making the sex cells to be deposited in their eggs they draw alternately from either one. If there is a unique gene in one set that does not appear in the other, it will be handed down to every other offspring.

The result is that there is a 50 percent chance that any two sisters will have identical copies of some gene handed down by their mother and a 100 percent chance that they will have identical copies of those handed down by their father. Adding these up, we see that any two sisters will be related by $\frac{1}{2} + \frac{1}{4} = \frac{3}{4}$ (or $100 + 50 = 150$).

By contrast a sister will share with her brother none of the genes inherited from her father, and only half those inherited from their mother. Their degree of relatedness is $0 + \frac{1}{4} = \frac{1}{4}$ (or $0 + 50 = 50$).

scrub jay. He found that out of seventy-four cases, helpers were assisting their own parents forty-eight times, a father and a stepmother sixteen times, a mother and a stepfather twice, a brother and his mate seven times, and an unrelated pair only once. Having the benefit of a helper's time and attention increased the survival rate of chicks hatched in the nest by nearly three-fold. The helpers did not lay eggs; their main contribution seemed to be in detecting and mobbing predatorial tree snakes. A kin selectionist interpretation of this behavior is that it is adaptive because the tree snakes make establishing a nest without a "guard" a chancy business. A young bird will reproduce better by staying with its parents and helping them make more of its brothers and sisters than by trying to initiate a nest on its own. If the tree snakes were to be wiped out, then the balance of benefits would reverse and the quality of familial devotion in the scrub jay would suffer. Kin selection nas also been invoked to explain the practise, among ground squirrels (*Spermophilus beldingi*), of giving an alarm call when a predator appears. Paul Sherman, of the University of California at Berkeley, was part of a team watching a colony of these squirrels respond to predators (weasels, badgers, dogs, coyotes, and pine martens). He found, first, that voicing an alarm call was dangerous; the alarm givers were stalked or chased by all five groups of predators more often than noncallers. But a squirrel was more likely to incur this risk if it had kin in the area.

Kin selection is immediately plausible to a lay human being, but it is still controversial among professional biologists, many of whom doubt that even the prospect of benefiting close relatives would necessarily cause an animal to select charitable and generous behavior. For instance, some ornithologists doubt that scrub-jay helpers get enough reproductive success from being "aunts" and "uncles" to make their behavior adaptive. These biologists believe that the young birds stay at their relatives' nests because good spots to make scrub-jay nests are very limited, and the helpers hope to inherit their relatives' territory. Of course the young birds are also rewarded for their altruism through the breeding of more kin, but more important is that the helpers have improved their prospects of being able to raise their own family. One reason these ideas are advanced is that the helpers do not work very hard; the birds themselves do not act as though the major reason for their being is to make more kin. What they seem to be doing is performing just enough work to compensate their parents for the nuisance of being underfoot and competing for food all the time, so that their parents won't drive them off. The parents, of course, have an interest in one of their own progeny's inheriting the property, so they will let the young birds get away with less work. Thus there are at present (at least) two explanations as to why young jays help out at relatives' nests: First, because whatever effort they do expend gets used in making kin, and second, since relatives have a reason to be more tolerant of them, the helpers will be able to get what they want (residency) with less work. If they tried to help at the nest of an unrelated jay, the young birds would have to strike up a true symbiotic relationship, a business deal, in which they gave as much value as they received. Why should they bother with what might be real work when they can get by with just loafing around at home?[94]

Animal societies are not always such a picture of striving, pushy, self-promoting individuals; occasionally one runs across observations that seem to be virtually Christian parables. One of these was contributed by Anne Rasa of the Max Planck Institute in West Germany. She was doing a long-term study on mongoose society when one of her males (Male 5) contracted kidney disease, weakened rapidly over a month, and then died. Mongooses are monogamous mammals (one of the very few), and live in groups of pairs organized in a dominance hierarchy. Male 5 was fairly low-ranking, but, as his illness progressed, the band allowed him priority access to food. Further, while before his illness he was only rarely in physical contact with the top couple, when he became sick the top-ranking pair increased their periods of contact with him tenfold, and, toward the end, spent the whole of their resting periods in contact with him, grooming him more and more, and licking him as his efforts to stay clean failed. Ms. Rasa believes that her mongooses were so solicitous toward the failing male because a larger band can mount a more vigilant defense against birds of prey, the mongooses' major enemy in the wild.

To some this view of natural relationships may paint a depressing amount of self-centeredness. I would be the first to admit that there are other ways of talking about natural societies; over the centuries there has been a regular alternation of descriptions that stress social cooperativeness in nature and those that emphasize individualistic self-advancement, and no doubt these zigs and zags of fashion will continue for centuries yet. But for the purpose at hand, for investigating the behavior associated with the evolution of sexuality, this emphasis seems like the right one. Compared to sexual creatures, asexual or less-sexual organisms have achieved higher degrees of social discipline, more elaborate organizations of mutually dependent specialists, and more striking examples of self-sacrifice and altruism. Sexual creatures have developed more elaborate ways of leading an independent and self-sufficient life while competing with other animals with similar needs. There is an ironical aspect to this: Sex requires mating, which is a very intimate activity. In other words the very adaptation that animals have turned to to help them compete with other members of their species requires them to cooperate with their competitors, at least briefly. While sexual reproduction seems to promote independence, it certainly also makes sexual creatures dependent on one another for gene transfer. The question of how these issues of cooperation and dependency are resolved in a context of strongly independent individuals is an important one, and it is with this question that the rest of this book is concerned.

CHAPTER

VII

THE EVOLUTION OF GENDER

THE ARGUMENT SO FAR RUNS AS FOLLOWS: SEX—GENE-MIXING—evolves when the members of a species find that their lives' outcome depends on how well they respond to social competition. These contests are usually not directly competitive, in the push-versus-shove sense; the affected animals may never see each other. Rather they are competitions of skill, competence, and efficiency. The animals that reach the important resource first do so not because they fought for it, or out of good luck, but because they were more of a master of the conditions at hand than other members of their species.

The more important this sort of indirect competition becomes the more likely it is that specialists will evolve. Animals will appear that are better equipped to deal with some of the situations they might encounter, and less well suited to deal with others, than a generalist would be. When this happens sexual reproduction becomes adaptive, since it allows parents to have, as it were, a "complete line" of progeny; some offspring out of the "line," ideally, will turn out to have the equipment required for whatever specialization has the most openings, the best opportunities, for the time and place at which the progeny become self-supporting.

However, evolution is not just a matter of animals responding only to their outside environment. Tools can remake their users, and often do. Sex may have been selected as a way of allowing parents to compete more effectively with other parents, but sex itself influences the nature of the creatures that use it. It makes them more self-sufficient and independent, and enhances those qualities we refer to when we speak of animals' being "wild and free."

So sex is selected in the first place when the social context in which animals live becomes more competitive, and to the extent that it has any independent influence on the nature of those who use it, it makes them even more competitive and independent. Yet at the same time, as paradoxical as it seems, it is impossible to have sex, gene-mixing, without an act—mating—that often requires a profound degree of intimacy, cooperation, and mutual vulnerability. Many issues need to be resolved before two wild animals can mate—where, when, who, and how, among others—and for each one of these issues the same question arises: Who is going to bear the burden of the resolution? Who is going to be inconvenienced most? Every animal would prefer to mate at a place and time that best preserved the order of its daily routine. (The very fact that these creatures are sexual at all suggests that they are specialists at what they do, and therefore especially likely to suffer from diversions.) Some animals live in multisexual colonies or as mated pairs, and for these where and when may not be a problem. (There are a number of species of anchored marine invertebrates, for instance, in which the sexes live intermixed. Fertilization is accomplished through a simultaneous, massive volley of sex cells, in which both eggs and sperm are discharged into the water. The discharge is triggered by a specific environmental signal, such as the angle of light penetrating the water, to which both genders react in the same way at the same moment.) But most animals live sufficiently dispersed so that some searching is done by one or both mates prior to mating. In all such cases there is a question as to who is going to bear the burden of the search.

The mating interaction imposes other burdens as well. Assurances might need to be made as whether the prospective mates are appropriate for each other, whether, on the most basic level, they are in fact adults of the right gender and species. Someone is going to have to get that information together and transmit it clearly, quickly, and credibly to the other mate. There may be a surplus of mates and some point in choosing among them. If so, then there might be an issue as to whether the mate in demand is going to have to evaluate the several candidates and choose one or whether the surplus mates will have this burden imposed on them as well, in which case they will have to select a champion by themselves and then send the winner on. Finally there is a question as to how the labor of getting into the reproductive state is going to be divided. For many animals the transition from foraging to breeding is a drastic one, involving many changes in behavior and physiology. It would be convenient to avoid putting these changes off as long as possible, that is, until the prospect of being mated was both imminent and certain. The ideal for any one animal would be to get so much attention from its mate that it could synchronize the transition into the reproductive state by those efforts alone.

If all these issues were resolved in favor of the same animal, then when it was most convenient for that party, a partner would appear exactly where that mate happened to be; the partner would communicate whatever information was of interest, and would set up a beat or rhythm of some sort that would allow a fast, synchronized transition into reproductivity. If more than one such mate appeared, then the candidates would settle among themselves the question as to which one best fits the mating requirements of the favored mate, and send on only that individual.

The remaining pages of this book argue that this is fairly close, more often than not, to the deal that females actually do get in nature. When a female moth wishes to mate she releases a tiny quantity of pheromone and all the downwind males within hundreds of yards begin flying toward her. Female salmon choose, for some reason, to mate in a place that is enormously inconvenient for a fish that lives in the sea; but whatever other problems this habit might make for her, acquiring a mate is not one of them. Whatever stresses and strains are involved, male salmon will follow her wherever she goes. The males of a number of species of mammals, including giraffes, routinely taste their females' urine in order to monitor her estrus cycle and be able to mate with her when she is ready—a behavior which guarantees these females a very efficient, convenient fertilization.

The reason males behave this way was mentioned in the first chapter: they have a problem with competition. What a female does to reproduce: build up eggs and superintend their development, takes time; what a male does—find fertile females and fertilize them—can take next to no time at all (depending on the social situation). This means that males will usually find themselves reentering the breeding pool much faster than females, and that a population of competing males will arise. In such cases females will have a choice of suitors; naturally they will use this leverage to pick the male that makes matters most convenient for them—the male who is where they want, when they want, and who behaves properly in other ways as well. The fireflies that one sees flitting and flashing about in the evening are usually males; the females are carefully sequestered away in the underbrush, sheltered (one assumes) from nocturnal predators like bats. Meanwhile the male has to fly around advertising himself in a most conspicuous manner. When a female catches sight of a flash pattern that promises a male of her own species, then, and only then, does she respond with an answering flash (often weaker than the male's). As mentioned earlier, there are firefly predators that mimic these flashes and capture and eat the males that land in response. If it were the females that were flying around looking for males, then they would be the ones to run all these risks. But males have to run them, because they have competitors; if they play a cautious, careful waiting game, then all the females in the area will be fertilized by males who are willing to run the risks of mating with the females on the females' own terms.

It may be that males exist, as males, only in order to cater to female preferences. In the first chapter the question was raised as to how creatures as apparently wasteful of their energies as males could ever have evolved. After all, there are other sexual reproductive systems in which all the members of the species are clearly involved with the manufacture of offspring, the simplest of which is bisexuality, in which everyone makes both sperm and eggs. How could a system in which half the species abstains from making this sort of direct contribution ever establish itself against the competition of these alternatives?

One answer might be that females who retain control over the circumstances of their matings can use that control to become much more productive and efficient. In these cases females that hired an assistant who did nothing but minister to their reproductive convenience might outcompete bisexuals who tried both to accumulate protein for the eggs and to give and get fertilizations. There are a number of ways in which having males as go-fers of the mating interaction might make females more effective. One is that females are, as stated, specialists, and a system of competitive males would allow them to continue to pursue their specialization, whatever it might be, as long as possible. A second possibility, which by no means excludes the first, is that competing males give females a very useful degree of control over their own reproductive schedules. There are two schedules involved: the development of the fertilized egg, to the point of release, and the packaging of progeny sets over the female's entire life-span, so that she can have a large number of litters, or clutches, or whatever.

An asexual female has total control over both these schedules and can adjust them to take maximum advantage of whatever possibilities there are in the environment and her physiology. But a sexual organism has become dependent on an external factor: the arrival of other genes. A delay in their arrival, or a general lack of predictability on when they might arrive, could prevent a female from keeping either of these schedules from running at their optimal rates. It is a mistake to think that the egg sleeps away in the female like a princess in a fairy tale, waiting only to be awakened by the kiss of the sperm. By the time the sperm first appear, each egg has had a long history; it has been constructing the biochemical equivalents of machine tools and assembly lines, stockpiling subassemblies, preparing energy sources, and generally tooling up and gearing up for the long developmental process ahead. It is true when the sperm arrives that the cellular divisions and differentiations begin; but none of this activity has anything to do with the genetic material brought by the sperm—that is not unpackaged for hours. Dividing and differentiating is something that the egg does by itself, at least in the beginning, in response to a sperm merely having arrived. There are a handful of species (nematodes, planarians, earthworms, minnows, goldfish, and salamanders) in which the sperm does nothing more than this; its genes are never expressed in the offspring at all. In other species the influence of all the prefertilization activity extends throughout the whole life of the offspring. A good example is that of a snail whose shell can spiral either clockwise or counterclockwise; which direction the shell of any specific snail will curl is already a settled issue by the time the sperm arrives. Biologists can induce the eggs of a number of species, including frogs and mice, to develop into perfectly healthy adults without any fertilization at all. In short, the relationship of an egg to a sperm is not that of a seed waiting quietly for its sun and water, but that of a busy industrial executive with a packed schedule, for whom the arrival of the sperm is just one event, even if an important one, among many. For such an executive punctuality is important, and a competing-male system forces the males to be punctual, since only the punctual males get to reproduce.

There are many natural systems that fit into this female-service idea of male purpose. One is a marine worm (*Grubea clavata*) that reproduces via the following minuet. First the females build their eggs up to a state of readiness. Then they advertise that readiness by ejecting a little squirt of egg protein into the water. This squirt alerts the males, who come closer and respond with a small sperm ejaculation. The females inspect this ejaculation to see if it belongs to the right kind of male; if it does, they release their eggs. This act stimulates the males to a massive sperm ejaculation. The point is that the whole process was initiated by the females, even if the first step they took was a subtle one; it was the females that defined the moment when the mating interaction took place.

An especially persuasive dramatization of the idea that female convenience has shaped the evolution of the male gender is found in the anglerfish. These are a group of deep-sea dwellers, about one hundred species strong, who get their name from the foraging practises of some of the females, who dangle a little phosphorescent lure in front of their mouths to attract prey. In many species of this group the males lose their jaw teeth early in their youth and thereafter draw their nourishment entirely from internal food stores. They replace their jaw teeth with a set of pincers (whose purpose has nothing to do with food) and set off to search for females; if they do not find a female before their food stores are exhausted, they will die. The males have enormous nostrils—as much as one-quarter of their head—and extremely well-developed eyes for an animal living at depths where no sunlight ever penetrates. They have been called "sexual guided missiles," but are really more like sperm with sensory and transport systems. If and when they find a female they lock their pincers onto whatever part of her body they bump up against. The circulatory systems and skin surfaces of both sexes merge. The male loses his eyes and fins. The two become, literally, of one flesh.

At this point, collected specimens seem to show, the female, who is already substantially larger than the male, begins to accelerate her growth, becoming, sometimes, a hundred times his size. The male also grows a little, drawing nourishment from the female, and enters into sexual maturity. The female develops her eggs. The male-female merger allows the male to monitor the female so that when she ejects her egg case into the water the male's sperm follow immediately. The egg case, which is the size and texture of a kitchen sponge and contains hundreds of thousands of eggs, begins to swell and suck in water, and with the water, sperm. It floats to the surface where the fertilized eggs hatch and feed. Eventually the weight of the developing young draws them back down to their parents' world.

Ichthyologists call these males *parasitic males*, and no doubt when a female is first seen with one (or sometimes more) sticking out of her the resemblance to a parasitic lamprey dangling from a lake trout is striking. But the resemblance is only a casual one. The eels get no help from their trout hosts. The angler females advertise for males, both by releasing pheromones and, judging from the size of the male's eyes, by some kind of visual signal of which we know nothing. Then they accept the male into their own tissues. Any immunologist who has been struggling to make it possible for humans to exchange simple skin grafts will appreciate what must go into an ability to combine separate, unrelated individuals. (Another bit of physiological acrobatics tossed off by the angler is that even though the circulatory systems are linked, the sex hormones of the female do not affect the male testes—nor the male hormones the female ovaries—adversely.)

The angler lives in a habitat with no landmarks—not even a horizon. Food is not plentiful; the population density is low; females are widely dispersed. It is difficult to find anything in any given period. In these conditions the female anglers have solved the problem of being mated at the right times by evolving a system in which each female has her own little toy, pocket-sized male, which she keeps with her and supports, and who in return has sex with her when and where she specifies. The angler male frees the female from the need of roaming around the ocean searching for males and allows her to pursue her notably low-energy form of foraging, in which she waits patiently for her prey to come to her, without sexual distractions.

The anglerfish illustrate how important it can be for a female to have a male on her terms, and not to have her own routines disrupted by the need to hunt up a mate herself. Every species, and group of species, has its peculiarities, however, and the anglers are not representative of the usual case. As a rule males conform to female convenience because their speedier mode of reproduction builds up a population of competing males that gives females a choice of mates. When some issue involving the mating interaction arises, the females will pick the best and/or most convenient candidate. Thus those males that conform to female preferences will have greater reproductive success than those that don't. The anglerfish are monogamous, though, in the most extreme sense possible. There is probably no population of competing males driving the system here. In fact males are rare enough so that an angler female does not, apparently, even commence her final growth surge until she already has a male in her pocket. She waits for the arrival of the males rather than tuning her growth rate solely by the possibilities of the environment, and when one does finally appear, she supports him. It is very rare for females to go to that kind of trouble for a male.

When one looks at the problem from another angle—the distribution of bisexual versus gendered species—the theory that males evolved for the convenience of females seems to pick up some support. In general bisexual species are found among snails, sponges, slugs, and worms, creatures that acquire their food by filter-feeding, continuously processing a relatively low-quality intake. Bisexuality is not found in species that look for isolated little bits of high-quality food, species for whom the issue of getting there first is important. Bisexuality is also found in environments that support low-population densities: in the higher latitudes, the deeper zones of the ocean, and the fringes of the richer habitats. This phenomenon is well-enough documented to have its own name—*spanadry*, the decline in the proportion of males toward the borders of a species' range. American eel males are more likely to be found in southern waters than in northern. Male-producing species of copepoda (copepoda are the marine insects, tiny, enormously various and abundant, herbivorous invertebrates) fall off regularly as greater depths of water are sampled. In one surveyed millipede (*Polyxenus lagurus*), the fraction of males fell from 40 percent in France to zero percent in Finland, with a steadily falling gradient in between.

If males evolved to give females a finer degree of control over their mating circumstances, then one would expect to see bisexuality where and when this degree of precision was less important. In general, a survey of bisexual species does seem to point in that direction. When the production of males falls off, it seems to be where females are less often affected by each other, either because population densities are low, or because their system of foraging is inherently noncompetitive.

A common explanation for bisexuality is that it evolves where it is important not to miss any mating opportunities. On the odds, only half the encounters between two gendered individuals will be between potential mates; any two bisexuals can mate.* Since bisexuals tend to live in sparsely populated habitats or are slow-moving, it is more important for them to convert every encounter into a mating. Darwin himself devised this theory in a famous work on barnacles. This is a plausible answer to the question: why bisexuality instead of sexuality? But it is possible to turn the question around and ask it differently: why sexuality instead of bisexuality? No doubt it is useful to bisexuals that every meeting be a possible mating, but why shouldn't all creatures be endowed with that flexibility? Why narrow the range of candidates at all? No doubt it is an advantage to (some) bisexuals to be able, in extremis, to fertilize themselves and start a new colony single-handed. But shouldn't we all be able to do this? Having a system of males closes down these options; what are the compensating advantages? One answer to this question is that when animals find themselves living in a highly competitive environment the reproductive cycle splits into

* As Woody Allen has pointed out, bisexuality doubles the chance of getting a date on Saturday night.

two subspecializations: a manufacturing and a service aspect. Females are what we call the animals that specialize in manufacturing; males, those that specialize in service. Putting it still another way, males have been devised by females to aid them in their competition with other females.

CHAPTER

VIII

THE PRISONER OF GENDER— AN UNRESOLVED MYSTERY

SO FAR ALL THE IMPORTANT QUESTIONS HAVE HAD ANSWERS, albeit, in most cases, speculative ones. But there is one feature of gender which seems inexplicable to me, and for which no satisfactory explanation seems to exist in the literature. This is the fact that most vertebrates seem to allow their parents to determine their gender, at conception, rather than making that decision themselves later on in life. By contrast there are a great many invertebrates that can switch from one gender to the other, depending on where the advantages are.

Gender flexibility has many benefits. For example, making sperm takes less energy than making eggs, so when several species of worms are starved they become male; when fed, they switch back to being female. (There are two genera of orchids—*Cycnoches* and *Catasetum*—in which flowers that grow in the shade become male while those that grow in the sun become female.)[30b] When there is an ecological advantage to being one sex or the other, these creatures can follow that advantage. Sometimes the benefits of gender-switching might spring from the social circumstances, such as when there are many females in the area it is better to be a male. There are a great many invertebrates—sponges, parasites, mollusks—

that have just this ability of assessing the social situation and choosing their gender accordingly. In one experiment, when the young of a certain marine snail (*Crepidula*) were isolated in an aquarium, 30 percent in one sample and 10 percent of a second began making sperm. When they were put in an aquarium with females of their species, 90 percent of both samples became males. The longer the association with females was maintained, the more time the snails spent as males. The experimenter concluded that the important physiological stimulus was the tactile sensation received by each young snail on his penis. When his penis stopped being stimulated, then, and only then, did he switch over and become a female.[118] The advantages of this flexibility in strict Darwinian terms, let alone its common-sense appeal, are so obvious that I cannot imagine why the vertebrates, by and large, have lost this ability.

Whatever the reason is, gender rigidity is not a necessary part of being a vertebrate, because there are a few that retain this power. One is the bluehead wrasse, a fish that lives on Caribbean reefs. The species is divided into two size classes: small fish, of both sexes, with variable coloration; and larger fish, all of which are male. These latter are very distinctly colored, with a blue head, green body, and two prominent black bars separating the two colors. The large males set up territories near the "leeward" side of the reefs, where, when eggs are released, the currents carry them away from predators that live on the reef. Spawning occurs daily, throughout the year, around midday. "At this time," a report on this species says, "most of the sexually active fish gather at a specific site on the reef. There the larger bluehead males set up temporary spawning territories along the outer rim of the reef while the smaller . . . males gather just inshore, . . . often massing in the hundreds."[150] The females, which can spawn every day, attempt to mate with the larger males, sometimes even waiting in line if a large male is occupied with other females. Fertilization is external, and stimulated by a "spawning rush" in which the male and female release their sex cells simultaneously at the apex of a rush toward the surface. While the smaller males do get some mating opportunities, the large males mate far more frequently, often fertilizing as many as forty times a day, and sometimes, one hundred. Thus being a large male is a much more attractive line of work than being either a small male or a small female, and, when bluehead wrasse females get large, some of them do indeed switch sexes and try to establish a territory of their own. When a dominant dies, or is removed by a curious ichthyologist, all the large members of either sex compete on a more or less equal basis to replace him.

A related species, the cleaner wrasse (*Labroides dimidiatus*), makes its living by cleaning parasites from the jaws and gills of a wide range of other fish that visit permanent cleaning stations maintained by the wrasse. (Fish swim from out of sight right to a specific area where they stop and wait.) The social unit consists of a single male, a group of three to six mature females, all of whom he fertilizes, and several immature wrasses of no developed gender, all living at a single station, or territory. The largest, oldest individual is the male; he gets to clean whom he wants. The largest female lives in the center of the territory with the other females scattered around her in various subservient positions. If there are two equally large "top" females, then the territory might be divided between them. The male defends his territory against neighboring males and frequently visits the cleaning sites of the females, who are comparatively more sedentary. There he feeds and interacts aggressively with them. "The male is more aggressive," writes D. R. Robertson, a student of these wrasses, "toward those females most likely to change sex and threaten his position—that is, larger females, especially the dominant one." Robertson notes that there is a special aggressive display, not seen in either female-female or male-male interactions that is employed specifically by males against the larger females.

When the male dies there is a general rush to take over his territory between neighboring males and the largest resident female, who tries to change sex and become a male herself. If the group is one of those with two dominant females, they might each change into males and then divide the territory and the "harem" of other females between themselves.

> For approximately half an hour after the death of the male the dominant female continues to behave . . . as a normal female. . . . Approximately one and a half to two hours after male death, maleness appears in the form of the special male aggressive display that the new "male" starts performing to the females of its group. The assumption of the male aggressive role can be virtually completed within several hours, when the "male" starts visiting its females and territory borders. The switchover to male courtship and spawning behavior . . . can be partly accomplished within one day and completed within two to four days.

The virtues of gender-switching are not confined only to giving females a crack at being male. There are at least two species of coral fish (*Amphiprion bicinctus* and *A. alkallopisos*) in which males seem to contend among each other for the advantage of being female. These fish, also called clown fish because of their variegated coloration, live symbiotically with sea anemones. The clown fish are immune to the anemone sting, and when predators threaten they retreat behind a shield of tentacles; at least in some species, the clown fish are known to share food with the anemones. The fish are large enough so that, as a rule, only two adults, a male and a female, can live on each host. The two fish seldom change partners, perhaps from the risks involved in swimming to another anemone. The typical social unit consists of a large female, a single smaller male, and a varying number of even smaller subadults and stunted juveniles, none of whom are offspring of the adult pair. The female dominates the male and they both dominate the smaller fish.

Specifically, the female regularly attacks the two top-ranking males; the ranking male focuses his energies on the second- and third-ranking subadults, and so on. Each fish attacks most vigorously the one immediately below him or her in the hierarchy. This "intragroup social pressure," write Hans and Simone Fricke, who studied these fish for three years in the Red Sea, "determines the gonadal development of the subdominants. Their testes are smaller and show little or no testicular tissue. Low-ranking males are psychophysiologically castrated."

When the females were removed from twenty-four clownfish pairs, eighteen of the males turned into females and recruited new male mates (of unknown origin). Some began laying eggs in less than a month. When different-sized males were paired experimentally, it was always the largest and more dominant that changed into a female. Interestingly, the females refused to change back again; pairing of females resulted in the death or serious injury of one of them, not in a switch to masculinity and subordination.

The Frickes explain this system by arguing that everyone is competing to be a female. The dominant female prevents the production of more females by actively suppressing those males that are the most likely candidates for gender-switching, while those males suppress other males that threaten to get ahead of them in the queue to be an egg layer. The Frickes do not say what the advantage is in being a female, or why the top-ranking female should care one way or the other what gender the second-ranking clown fish chose for itself. One might imagine that the second-ranking male would be overjoyed if the third-ranking fish were to choose to be a female; this would give him two females to fertilize. What would be a real problem for him would be if male three stopped being "psychophysiologically castrated" and started fertilizing the group female instead of him. Another interpretation might be that all the fish are really competing over the most secure spot inside the anemone's tentacles. The larger fish takes the best place; the second larger the second best, and so on. Each fish then adopts the gender that fits its life situation best. The largest fish becomes a female because, being largest, it can make more eggs than anyone else. If it stayed a male and let one of the smaller fishes become a female, then fewer eggs and less procreativity for itself would result. The second fish stays a male for the same reason. He has no reason to be a female; he is already a parent of all the eggs produced by the pair. Switching gender is not going to change that, and since he is smaller to begin with, he will lose progeny, not gain any. It is true the female harasses him, but that might be to remind him to stay out of her place, not to keep his gender. And the little fish stay sexless and small because, if they tried to be male, the resident male would harass them to death by driving them away. (Conceivably they might try to be very small females.)

The point is that in many species creatures have the capacity to follow opportunity from whichever gender it beckons. The advantage of being able to fine-tune one's sexuality to the conditions immediately at hand seems obvious. It is very common in sexual species for a significant fraction of the males to be shut completely out of all mating opportunities. Why don't those males then try to become females, no matter how incompetent and inexperienced? And yet there are very few gender-switching vertebrates and no warm-blooded animals at all with the knack. Gender in all these species is determined at conception. It apparently is decided permanently by the parents alone, and by and large, for whatever reason, vertebrate progeny seem to have allowed this important decision to be made for them.

CHAPTER

DOING THE PAPERWORK AND OTHER ISSUES

MANY ISSUES HAVE TO BE SETTLED AS MATING UNFOLDS, AND, in part because it is such a departure from the ecological, physiological, and social routines of everyday, the resolution of each one of these issues might demand a considerable amount of inconvenience, stress, and risk. Two such issues are time and place. Often all the members of both genders live scattered over the landscape; in other cases the females live in groups but the males forage elsewhere (sometimes because the females actively repel any male horning in on their territory outside of breeding season).

A third issue is who—if a choice is available, which one should be the mate? A mate might be preferable because he (or, for that matter, she) makes the mating emotionally and physically convenient and comfortable; because he or she has control of a valuable resource; or because he or she has an intrinsic, personal attractiveness. In the terms used here, this last means a genetic superiority, an inheritable advantage, and that advantage, in turn, might speak to an enhanced ecological, or physiological, or social adaptiveness. (Many biologists believe that an intrinsic genetic superiority is the most important issue of all in mate choice.)

These issues require that information be exchanged. Somehow the partners have to learn where to meet when the female is receptive. If a choice of mates is to be made, the chooser has to acquire enough of the right kind of information to make a sensible decision. On the simplest level, it is important that mating interactions involve two creatures of the same species and the right maturity level but of opposite genders; otherwise they are absolutely certain to be a complete waste of time. Other values can enter into the mating choice as well. Somehow all this information has to be gathered, translated into transmissible form, then transmitted, received, and decoded. This can be a lot of trouble.

If one looked at the question of "paperwork" abstracted from the realities of sexual politics, one would conclude that females ought to be the ones to shoulder the burden of the paperwork because the issue is more important to them. A female places a larger fraction of her reproductive energies at risk in a single mating than does a male. Fertilizations for a male are (once a female has been found) cheap and fast. A male will seldom find that he has sacrificed an opportunity to fertilize one female or females because he took too long, or exhausted himself, fertilizing others. Each new female is therefore all gain; there is much less need to weigh alternatives and make choices. The opposite is true for females. Therefore what one might expect is that once two mates came together, the female would actively explore the male, subjecting him to a searching physical and mental examination, while the male, in his turn, would demand fewer tests of the female.

Half of this prediction is confirmed. Males are, in fact, less discriminating than females. In some insect species, males can even be seen courting females of the wrong species. (Some male flies have been seen mating with raisins.) Male frogs and toads, particularly in those species that breed in aggregations, clasp male and female alike. Male elephant seals mount females whether they are pregnant, giving birth, postpartal, or fertile. There are even a number of species of orchids that pollinate themselves by "parasitizing" the sexual indiscriminativity of male wasps. This phenomenon was discovered in the late nineteenth century when a French botanist living in Algeria, A. Pouyanne, noticed that the flowers of one orchid species seemed to be visited only by the *males* of a certain wasp species (*Scolia ciliata*); the females were completely indifferent. When Pouyanne investigated, he was so startled by what he found that he accumulated observations for twenty years before he published a word. Eventually he and other botanists showed conclusively that in this and other cases, the males copulated while in the flower and received no food. These facts and the rough visual similarity between the flower and the female wasps convinced the investigators that the orchids imitated both the odor and appearance of the females. The orchids thus enticed the males into a bout of sexual activity that ended by the males' carrying off pollen and therefore advancing the sexual ambitions of the orchids. "It may be that those who would reject the evolutionary approach to an understanding of life," wrote a botanist of these flowers in 1937, "and who prefer to regard the world as the product of Special Creation, will lean a little more lightly on human weakness when they discover moral turpitude among the insects."[5] Once the female wasps have emerged in large numbers, incidentally, the number of "parasitic" copulations declines rapidly.

Males are less discriminating than females; this much is true. But when one examines a representative range of mating interactions, one does not see the female busily checking over a male, examining his good points and bad points like a buyer at a horse auction. Instead, as a rule, the females simply sit there, often looking elaborately uninterested, while the male engages in a colorful, complex, and raucous communication called a courtship display.

Quite a number of these displays have already been mentioned: the sumptuous hut of the bowerbird, the complicated dance of American grasshoppers, the bands of turkey brothers, pinwheeling in synchrony around the female. The list could be extended indefinitely. Fiddler crabs wave their claws; at least one species of water strider (a *Rhagadotarsus*) courts through the propagation of surface vibrations; male *Sceloporous* lizards do push-up displays; male tortoises nip the female and ram their shell against her. There are all the sounds of spring: birds, and crickets, and tomcats.

We ought to look at one display in a little detail; perhaps that of rabbits will serve. It has many elements. First, there is "courtship chasing," in which the buck pursues the doe back and forth across the field:

> . . . chases are interspersed with long intervals of false feeding, in the course of which both animals retain their alertness, and the buck will sometimes edge toward the doe until he is near enough to attempt to rush her.

They displayed "amatory behavior," such as licking and nuzzling:

> It is not uncommon for two rabbits to be lying face to face licking each other's muzzle and ears for a half hour or more. In most cases the doe then stretches out on the ground and the buck sits facing her, so that he may reach the back of her head and ears, which appear to be favorite places to be curried.

Another behavior was "tail flagging":

> In this commonly seen form of behavior the buck elevates his haunches so that he walks with a stiff-legged gait, and lays his tail flat along his back so as to display its white underside. . . . Various accompanying movements are performed, including "false retreat" in which the buck walks stiffly away from the doe for some six yards, giving her a full view of his elevated tail. He will then return to her and repeat the performance, sometimes three or four times in succession. Another accompanying movement is the "parade" in which the buck circles the doe at about two yards' distance, with tail elevated and his rear quarters twisted toward her. . . . The parade is not always circular, but is sometimes conducted back and forth along a line in front of the doe with the tail twisted at each passage in the appropriate direction.

The rabbits also practised "epuresis" or "enurination," in which the male (usually) squirts a jet of urine at his partner:

> The buck may merely turn his hind quarters toward the doe and shoot out a jet of urine backward, but some form of circling round the doe is more often involved, frequently preceded by tail flagging. A very common method is for the buck to run past the doe about a yard from her and to twist his hindquarters. Southern notes that it is remarkable how rarely these running shots fail to find their mark. A less common method is downward enurination, in which the buck leaps over the doe and emits a jet of urine as he passes over her.
>
> The effect on the doe is variable . . . but in a few instances definite stimulation is observed.

Who is getting what out of all this noise and fuss? The most widely accepted theory among biologists is that usually the males are filling in applications, telling the female, at a minimum, what their species, sex, and maturity level are, and perhaps are describing themselves in much more detail than that. One clue is that when related species live close together their courtship displays differ markedly. Some lizards court with an extensible chin pouch, called a dewlap, which is repeatedly and rapidly flashed in front of a female. On islands where lizards of the same size compete for similar resources, the males of different species never share the same dewlap color. For instance, the three following two-and-a-half-inch long species were all found living on the lower parts of trees in Camaguey, Cuba: *Anolis homolechis*, a dewlap of gray or white, and a body color of light tan with horizontal stripes and four dark chevrons; *A. allogus*, a dewlap of yellow to apricot with three to four reddish stripes and white margin and a body color of reddish-brown with yellow reticulations; and *A. sagrei*, a dewlap of bright red, dark red, or brownish-yellow against a body color of tan brown. Solitary species living on islands that can support just one or two lizard species generally all have yellow to yellow-orange dewlaps with gray-green body colors, even though these island lizards come from several different evolutionary lines and live on all kinds of islands (high and low; wet and dry, and so on). Exactly the same points have been made about the calls and plumage displays of birds. In its main habitat the Australian robin (*Petroica multicolor*) males are a bright red, white, and black, and the females a camouflage brown. In the small island populations, whatever was keeping this system disciplined collapses: hen-colored males and cock-colored females are often seen. The theory in both lizards and birds is that when there is a risk that the wrong sets of genes might get joined, the males carry unambiguous descriptions of their own genes and assert a sexual truth-in-labeling; when the risk ends, so does the need for elaborately error-free descriptions.

This explanation has gained power in particular from investigations into mimetic species, species that look alike (sometimes called sibling species). (In many cases the discovery that a habitat was divided among several look-alike species rather than a single one was made only when differences among the male displays were noticed.) For instance there was a case found in Haiti of three adjacent mimetic lizard species (all of which had been called *A. brevirostris*). The males of the northernmost species all had dark dewlaps; the males of the southernmost species, light dewlaps. But the males of the middle species had light dewlaps toward the northern part of their range and light ones toward the south. Their identities were emphasized most strongly where confusion was most likely. This might not prove the "gene-labeling" hypothesis, but it surely fits it nicely. In a series of experiments on two mimetic species of European grasshoppers (*Chorthippus brunneus* and *C. biguttulus*), it was found that in almost every case grasshoppers did not answer calls except to the opposite sex of their own species. The only way in which cross-matings could be induced was by having the songs of the right mates playing in the background. (In the wild, the male travels about, calling randomly, until a female answers. Then he shifts into a display song and approaches her. The two mate only after exchanging chirps for a while.) Sometimes experimental manipulation can expose the system even more clearly. In two ponds, each stocked with two species of sunfish, hybridization occurred only in that pond in which the male's display fins had been removed.

It is possible that females can use these "gene labels" to achieve quite a degree of fine-tuning in their mate choice. On Guadaloupe there is only one *Anolis*-type lizard, so there is no risk of two species cross-mating. Nonetheless the markings of these males are anything but dull. The species (*A. marmoratus*) divides into a series of races . . .

> . . . in one of which males have an apple-green body and a blue-gray head with brilliant orange marbling, while in another males are plain green in ground color but anteriorly and dorsally have dark blotches tending to run together and set off by pale cream borders. In yet another race males are pale gray-green with brown heads, and there are still other remarkably different patterns within this one species. . . .[159]

Some herpetologists put cases like this together with the observation that dewlap patterns and ecology seem to be associated (a male living in a wetter area might show more orange) and suggest that these lizard males have been selected to reveal in detail the kinds of environmental conditions to which their genes are best adapted.

Fruit fly (*Drosophila*) females seem to prefer, for whatever reason, "minority" males, the less common male types in a population. If five males from Texas are mixed with fifteen from California, the Texan males do better; but if the proportions reverse, so does the ratio of sexual success. (In the experiment just cited the females were divided equally between Texan and Californian strains.) Besides being able to pick out the less common of two strains, they can pick out the least common in more than two strains. This means that the females must (1) have a preexisting classification system into which the various males can be sorted, (2) count how many males there are in the sample that fall into each category, (3) compare and then rank the categories according to the number of males in each one, and (4) accept a male from the least-populated category. This might sound like a lot of work, but it does not compare with what the male is doing. He is generating the data, and to do so he uses all the channels of communication open to him: tactile, auditory, visual, and olfactory. He dances, he buzzes, he generates pheromones. In so doing he is laboriously filling out an application with all manner of detailed information about who he is and where he comes from. The female then evaluates his application. What the female does is cerebral and might sound more impressive, but it is really the male that has to burn the calories.

Why should the males do all this work when it is the females that stand to lose more from wasted matings? The answer, obviously, is that the females can afford to wait. While males run out of females before they run out of sex cells, females almost never run out of eggs before they run out of potential mates. Her eggs will be fertilized, if not now, then soon enough. If a superior male arrives on the scene, he will get around to her. The difference in reproductive tempi and the existence of the population of competing males give females the luxury of an extended contemplation of the issues.

But those males who can persuade females fastest that *they* are the ones the females want, who can settle reservations fastest, quiet misgivings most expeditiously, will suffer least from interference by competing males and have the most time to devote to scouting up additional mates. (In addition, in many species mating involves some risk to a female, either to predators, by exposing her or reducing her mobility, or to the hubbub of male fighting. A male that can establish his qualifications quickly might find himself being rewarded for that reason alone.)

Another issue of the mating interaction is who does the work of synchronizing the relation. When males and females mate they both, though to different degrees, have to abandon their accustomed habits and adopt new physiologies, behavior patterns, and social relations. Often they must both abandon an antipathy to a physical encounter with a creature of their own kind. An animal might be selected to manage these difficult transitions itself, or to help others pass through theirs, or both.

As a rule males are selected to manage their transitions on their own and to help females pass through theirs. Once again males find themselves adapted to conform to female requirements and conveniences, and the reason is the same: any single mating benefits the male more than it does the female. A male that brings a female into reproductive readiness gains a set of progeny he would not otherwise have had; progeny that would probably have been fathered by some other male had he not brought the female into receptivity. A female that brings a male into a sexually active state usually gains nothing but a mating that she would have gotten anyway just by waiting a little longer.

So it is not surprising that endocrinologists have found that male courtship also has a positive influence on the production of female hormones and general glandular changes. Experiments showing these effects have been carried out with canaries and other birds, reptiles, bullfrogs, and with many small mammals. (It has even been found that a female canary exposed to a richly varied, diverse, male song builds her nest faster and lays more eggs than does a female who has to put up with a more limited repertoire.) The more male courting behavior female lizards are exposed to, the faster their ovaries grow. It is the behavior, not just the physical presence of the male, that counts, since castrated males, who do not court, do not stimulate increased levels of hormone production.

There are a number of small mammals (cats, rats, rabbits) in which it is known that the mechanical stimulation of copulation, entirely apart from the actual ejaculation, is necessary for the female to ovulate. Copulation in rats, for instance, seems a rather cumbersome affair. The male mounts and dismounts the female over and over; then he intromits, or penetrates her, several times. Finally, after somewhere between five and eighteen thrusts, depending on the strain of rat examined, ejaculation occurs. Often then the male dismounts, waits several minutes, and then begins the whole procedure again. What the male is doing, with his mounts and dismounts, is building the female up to a level of stimulation such that she will enter into a receptive posture, called "lordosis". While in lordosis the female is, in the words of one investigator, "in a vegetative state". She becomes insensitive to many classes of sensory stimuli to which she would ordinarily pay attention. All her locomotive reflexes seem blocked. Her body is arched and rigid, and the pupils in her eyes dilate to their fullest extent.

The rationale of the repeated penetrations is found in the mechanics of the females' production of progesterone, the hormone that allows the fertilized egg to implant in the uterus wall (and performs many other pregnancy-related functions). Rat females do not produce this hormone continually; what switches production on are the repeated male intromissions. Females receiving too few intromissions from a male who nonetheless ejaculates do not transport his sperm from their vagina into their uterus. Males seem to be sensitive to the number of intromissions females need; in experimental matings between strains of rats that required different numbers of thrusts to switch on the progesterone, the males tended to conform to the females' needs rather than their own past practise.

A variant of the theory that male courtship helps females shift physiological gears is that male courtship overcomes the fight-or-flight reflexes so routinely relied upon by animals in their everyday life. This explanation has been explicitly advanced in connection with the intricate courtship ceremonies of the manakins, a group of about fifty-odd species of tropical fruit- and nectar-eating birds. The males of many of these species mount joint displays, involving two or more males in what is virtually a circus act.

Mercedes Foster, a zoologist from Berkeley, recently published a report on the behavior of the long-tailed manakin, *Chiroxiphia linearis*. In this species two, and on rare occasions three, males form long-term mutual attachments, persisting at least some of the time from year to year, over many breeding seasons, and independent of the chosen display site (if the display site is changed the birds move as a pair). The season begins by these pairs voicing a synchronous call, which Foster reproduces as "toledo". Unpaired males do not give this call and never attract females, and paired males almost never attract females except when producing this call. When a female arrives the birds . . .

> . . . move to a display site consisting of a low display perch and associated vines and branches. One male, or both alternately, performs a Solicitation Display which appears to stimulate the other male to proceed to a Jump Display. The Jump Display, which cannot be performed by one male alone, has a number of variants. The two most common are described here. In the Up-Down Variant the males perch perpendicular to the length of the display branch facing the female. In a coordinated fashion they alternately jump into the air, one bird hanging at the peak of the jump momentarily before descending to the perch. As he lands, the second male jumps, and so forth. As the male jumps, he gives a wheezy *buzzee* call. In the Cartwheel variant the males perch parallel to the length of the display branch facing the female located at one end. Again they jump alternately, but now the front individual jumps backward, landing on the spot previously occupied by the second male who moves forward to the anterior position. As long as the sequence lasts they make a continuous moving circle around each other.

After the joint display the top male of the pair (always the same one) performs a special, solo, precopulatory display in front of the female and then mates with her. The two or three males do not seem to be brothers, and Foster believes that the benefit to the subordinate male comes not through kin selection but because the subordinate is younger, will eventually inherit the display site, and will then become the dominant member of the new pair. The subordinate is thus a symbiotic apprentice, and is simply polishing his technique against the day when he will come into his own.

It is very hard to watch these displays and not try to imagine what purpose they serve and why they evolved. So far, the explanation that seems to have satisfied those ornithologists that have studied the manakins directly is that two males can, by displaying together in this coordinated fashion, overwhelm a female's normal feelings of aggression and fear at the prospect of a close approach by other manakins more rapidly than a single male could, plus they will attract more females as well.

What is there, one wonders, about a male jumping around that connects with a female's hormone system? Why do these displays work? The feeling I get, in listening to or watching the displays of local animals, is that they work by choosing sounds and motions that reproduce within the female the same feelings that are induced by good environmental conditions. In other words, imagine that females are selected to be fertilized in a certain range of temperatures, or when in the presence of a rich-looking territory. Such a female might be pictured as, at least on some days or in some territories, being a little uncertain as to whether the conditions of the moment are quite good enough. At such a point a male that loudly declares this to be the best day ever seen, or the most marvelously lush countryside imaginable, who induces in her the mood that those conditions would have created had they clearly existed might push her over into a positive decision. Certainly listening to birds court makes *me* feel happier, more expansive, confident, and ready for adventure.

CHAPTER

MALE SOCIETY—THE SOCIETY OF COMPETITION

IF ONE WERE BUILDING A MYTH AROUND THE NATURE OF males, the center of the drama would be the moment when females took over the reproductive cycle, monopolized it, and left males nothing to do but compete for their favor. When that happened females effectively emerged in control of male evolution, for it was only those males that met their terms and danced to their tune that were able to reproduce. Conceivably this need not have happened; had the males organized themselves, and agreed to ignore all the females, then eventually the females would have had to come looking for them. They might thus have gotten the females to do *all* the work, from accumulating the protein to arranging the mating (though actually if the males really had tried a trick like that the females might just have evolved a bisexual life-style and dispensed with males altogether).

But in reality gender solidarity was never an option; sexual creatures, as pointed out earlier, have an individualistic, anarchistic bent and seldom engage in compacts that entail sacrificing for the common good. (For example, there is no known case of a predator species—humans perhaps aside—successfully adopting a quota system, even though these can increase prey yields considerably.) Instead males compete with each other for the available females.

Such a competition might be organized in several ways. The most direct is simply sequestering the female; physically removing her from the attentions of other males. One example of this kind of competition was studied in a species of sand bee (*Centris pallida*) by the members of a zoology class at Arizona State University. The males of this desert dweller precede the females in hatching out of the soil by a few days. The females seem to dig upward and then pause just under the surface. The male bees cruise "in rapid, sinuous flight" about an inch over the desert. They seem to be able to sense alterations in the pattern of odors given off by the sand and often drop to the ground and walk about briefly, sweeping the soil with their antennae. Then they either take off again or begin gnawing and digging their way through the hard, sun-baked surface. Sometimes a male will dig down to a female in less than one and a half minutes; in other cases, he might need twenty. The average excavation took six.

When the male uncovers a female he begins to court her. This involves . . .

> . . . complex, rhythmic movements of the legs, abdomen, and antennae of the male and is associated with a low soft rattling auditory component, perhaps produced by the female. Males vigorously stroked the sides of their partners with their middle and hindlegs in a rhythmic fashion averaging about 1.3 strokes per second. . . . At the same time . . . the male's abdomen tapped the upper surface of the female's abdomen while the males antennae moved down along the female's antennae. . . .[3]

This courting behavior begins in the excavation pit but is usually interrupted so that the couple can fly to some nearby trees or shrubs; there courtship is resumed and at some point the female allows copulation to take place. The courting sequence takes only four minutes or so altogether.

The females mate only once and then fly away; the males remain, cruising rapidly over the surface, looking for female spoor. An intense male-male competition builds up rapidly. Digging males are often attacked by other males seeking to supplant them.

> Digging males defend their sites vigorously. When approached and touched by an intruder, the site owner will elevate his body (first backing out of the hole if it is one cm or so deep) and kick at his opponent. If the other bee does not leave, the digger will turn to face his antagonist and may then lunge forward. If the potential usurper stands his ground, the two will then usually rear up on their hind legs while facing one another. . . . If one bee is smaller than the other, he is likely to topple over backward and may leave promptly thereafter. Alternatively, the two may grasp one another and grapple violently, wrestling and twisting over the ground . . . until one manages to break away and escape.
>
> On the mornings of May 14-17, 1975, at the Saguaro Lake area, 50 records were made of activities at digging sites. . . . On 43 of the 50 records, a male was forced to interrupt his digging to repel an intruder by kicking, lunging, or rearing up at his opponent. A total of 393 such interruptions were noted or an average of 1.5 for every minute of digging. . . . Because the average time required to dig down to a female is about 6 minutes, males usually find it necessary to defend their digging sites repeatedly. . . . (Of the 50 sites and 393 attempts there were only 21 successful takeovers. In a second study area 10 digging sites were watched and four successful takeovers were seen.)[4]

The game, of course, is not over when the female is uncovered. Intruders are likely to challenge a successful excavator as soon as a female appears, to attempt to dislodge him during courtship, to interfere with his flight to the bushes, and to disrupt his attempts at copulation.

The female's response to all this scrapping and brawling is predictable enough: she flies away from it. If a male is distracted or dislodged by an intruder once the female has emerged, he might lose her regardless of his success vis-à-vis his competitor. The intensity of the fighting among the diggers (and the females' avoidance of it) makes possible a second male strategy, that of being a darter. Darters station themselves about one meter apart, on the margins of the regions where the females are emerging, near flowering shrubs and trees, and hover, looking upward. These males are smaller than the diggers and never make aggressive, physical contact with other male sand bees. Almost certainly they are waiting to spy females that have escaped the attentions of the diggers; in experiments involving the release of females such females were chased and "captured" in flight. The biologists believe that darters must have been doing the chasing, since the diggers seem to pay all their attention to the ground.

A direct, scrambling competition, in which females are fought over one at a time, is common in insects, but it is certainly not restricted to them. The common garter snake often overwinters in large aggregations. The males wake from hibernation first, crawl out of the cave or hollow, and then await, as a group, the females. The females wake over a range of several days and straggle out a few at a time. Each is courted by dozens, hundreds of males. Courtship in garter snakes involves a physical kneading of the female by the male; he rubs his head up and down over her back. A garter female, therefore, may find herself surrounded by a great many males all trying simultaneously to massage her. After a time she admits one of that number; a pheromone announcing intromission is released (it may be a mixture of substances produced by both sexes) and the mating ball dissolves.

A last example of direct competition for individual females might be the spawning practises of salmon. The female first builds the redd, or egg-laying site; then, according to some observations made in 1934:

> . . . she selects a favored male and will aid him in chasing away other amorous males, although she will not go far from the redd. In making her choice she favors the large males. . . . The role of favored male is not secure, since he may be supplanted at any time during the season, day, or hour. When there is a surplus of males on the spawning grounds, cruising males are constantly shifting from female to female, awaiting the crucial moment when they may share in the fertilization of the eggs.

> Within a period of five hours three different males may have been observed to occupy the position of favored lover, the larger cruising males driving off their smaller rivals. . . . When salmon are numerous on the spawning grounds, particularly at night, the water is kept in constant turmoil by the activities of the males chasing invaders from their chosen females. . . .[18]

A very common form of male competition is establishing territories—areas or resources to which receptive females will be attracted and from which males try to drive each other away. The competition is for the territories, not for the females directly, though there is sometimes an element of this, too. The males of many species of insects, fish, amphibians, and reptiles struggle over control of "ovipositing" or egg-laying sites upon which, once control is established, they wait for females to arrive. The dragonfly mating system is a classic instance of this. The females lay their eggs in lakes. The females of each species —and sometimes different kinds of females within the same species—have distinct preferences among the variety of textures (mud clumps, algal and other vegetation mats, sticks and logs, and floating objects of various kinds) a lake offers. The males usually appear first, fight over the sites that females of their kind of species will find attractive, and then wait for them. Similarly, African antelope such as the sable, the impala, and the waterbuck contend over fractions of the range on which females forage. Males mate with those females that happen to become receptive in the area he controls.

Most male mammals compete over "residency rights"—the right to stay near a group of females that will, at some point, come into estrus. A famous example is lions. The core of lion sociality is the pride, more specifically, the females of the pride. These are a related group (mothers, daughters, aunts, nieces, cousins, and sisters) of a half dozen or more lions, all of whom live together in a stable society in generally the same area for long periods of time (a lioness might live for eighteen years). A pride usually has two males as well. These are often brothers, but from another unrelated pride. They win their right to live with the females by driving out the previous set of males and usually manage to last for two to three years before being driven out in turn.

Another, exceptionally well-studied example is that of the yellow-bellied marmot, a chunky, squirrel-like creature known for its ability to live at high altitudes. Marmots are one of the more social of the rodents. Like prairie dogs (the most social rodent) they live in permanent colonies laid out over several acres and interconnected by an intricate series of home burrows, secondary or refuge burrows, and trails. Each colony is composed of two or more females and a male. (Prairie dog colonies are really supercolonies, made up of many such families, all of whom will join to conduct a group defense of their common territory against outsiders and predators.)

Marmots hibernate seven months out of the year; they emerge in May and stay active until early or mid-September. Happily this corresponds nicely with the activity cycle of most college students and professors as well, and this felicitous symmetry has given us an intimate acquaintance with marmot society. It echoes many of the features of lion social biology. The females are more likely than not to be the daughters of marmots that lived in that colony before them; the male comes from further away, is not related to the females, and is usually turned out after two to three years by a successor male. The males spend most of their time on the alert for challenges to their tenancy. Kenneth Armitage, a University of Kansas zoologist who has done more than anyone else to uncover and relate the intricacies of marmot society, has written:

> . . . The most striking characteristic of male adult behavior is its conspicuousness. As a colonial male moves around a locality, he carries his tail in an arc extending above and to the rear of his body. The tail is waved back and forth in a behavior described as flagging. . . . Flagging is so easily seen that it must serve to advertise the male's presence.[9]

Males either survey their territories from one or two look-points or patrol them throughout the morning. Flagging is directed at transient males that are not prepared to fight. Advertisement allows both parties to avoid the stress of a direct challenge. Sometimes a real challenge is made, though, and open fights break out. Armitage was lucky enough to witness one of these:

> 16 June 1968. 402 (the identifying number of the male) by the upper cabin. 355 (the resident) comes up past the wall. Both flagging and moving about. Much circling; first one advances, then the other. Each rubs head alongside bricks, etc., in the yard. 355 returns to the wall and 402 follows and marks bricks at the wall, then retreats to cabin as 355 follows. More circling and flagging. Each appears to expose the anal region to the other. 355 starts down bank past wall and 402 follows. Suddenly 402 jumps on 355 and vigorous wrestling follows. They appear to lock jaws. 402 seems dominant, then 355. 402 broke off and ran and 355 followed. 402 entered burrow under cabin porch and 355 remains at entrance, jumping and moving around and rubs side of head on boards above burrow entrance, then lies stretched out on board, then moves off.

The background of this particular blow-by-blow was that 355 was defending a combined territory, one which had "traditionally" been three separate male ranges. He did this successfully, despite repeated challenges, for four years, which is considerably longer than the average male residency in even single territories (two and a half years). Marmot 402 remained in the area despite losing this fight. He was trapped by Armitage's team the following year and introduced, as an experiment, into a second colony area. He left, and the next year, 1970, turned up in a third area where the resident male had died. Here he was finally able to settle in with a group of females of his own.

One of the more important elements of this fight is invisible to humans; 355 and 402 were almost certainly "shouting" at each other over the olfactory channel. The head-rubbing, the marking, the mutual exposure of the anal regions, all are probably connected with an effort to make the battlefield smell like one's own homeground and therefore make the opponent feel like a stranger, or at least less at home. Such odors serve the same purpose as strutting and swelling. They amplify the presence of their generator, or generatrix, and make his or her identity more pervasive. These odors, pheromones, are also used to control territory without physically occupying it; they serve the same purpose as birdsongs, or our system of property rights. Marmot 355, one might note in this connection, was challenged only on those portions of his extensive range that he visited least. He was never tested on that area where he actually lived, patrolled most often, and where, presumably, the sense (and scent) of his presence was strongest.

A third form of male-male competition—known in several species of insects, about thirty kinds of birds, some fish and frogs, and a handful of mammalian species—begins with what seems to be a statement of male organization. The males congregate and, acting collectively, pull the females in from the surrounding area. Only when the females appear do they compete, usually through displays. One of the most impressive examples is found among the East Asian fireflies. Enormous numbers of males assemble in hundred-yard stretches of mangrove trees at night along river banks. "Imagine a tree thirty-five to forty feet high thickly covered with small ovate leaves," an American biologist, Hugh M. Smith, once wrote of this phenomenon, "apparently with a firefly on every leaf and all the fireflies flashing in perfect unison at the rate of about three times in two seconds, the tree being in complete darkness between flashes. . . . Imagine a tenth of a mile of river front with an unbroken line of trees with fireflies on every leaf all flashing in synchronism, the insects on the trees at the ends of the line acting in perfect unison with those between. . . ."[26]

Biologists call these male mating aggregations *lek* systems. The lek itself is the area, usually a very small part of the habitat, that is used, often year after year, for mating displays. The fidelity of the creatures using the lek can be remarkable. One population of ruffs in Britain continued displaying on its traditional spot even after a road was built over it. The last surviving heath hen in Martha's Vineyard, Massachusetts, became the object of some local sentiment by persisting in returning to the display grounds of her species, year after year, and waiting fruitlessly for the booming, whooping, vocal displays she no doubt remembered from her chickhood.

When breeding time arrives, the local males migrate to the lek and crowd into a number of small, adjacent territories; from there they flash, whoop, boom, crackle, puff, flutter, and generally transform themselves into as dramatic and eye-catching a specimen as possible. The females wander around the lek, examining the various candidates. As a rule they end up all making the same selections. Usually only a tiny fraction of the males successfully mate; these mate with almost all the females. An example of a lek system in mammals might be that of the lechwe antelope (*Kobus leche* Gray) of Central Africa. Richard H. Schuster, of the University of Zambia, has described their display. During rutting he found a number of circular areas, about half a kilometer in diameter, within which fifty to one hundred males, spaced about fifteen meters apart, were displaying:

> The males rarely grazed, but instead stood in an exaggerated pose with head held high, legs prancing, tail wagging vigorously, and penis often erect. . . . displaying is enhanced by features present only in the larger adult male—lyrate horns (lyrelike) averaging seventy-two centimeters (two and a half feet) along the front curve and solid black markings on the fore- and hind-legs, with the foreleg patches extending up to conspicuous shoulder markings. [They also tear] at the ground with horns. The head is swung vigorously from side to side, causing tufts of grass to be flung high into the air and often draped on the horns and neck.
>
> . . . Lechwe leks were notable for the almost continual occurrence of male-to-male interactions between territorial occupants, conveying an atmosphere that was highly charged and unstable. The movement of a male or a female almost anywhere was likely to set off a chain reaction of chases and fights. Chases

> were more frequent. . . . Typically, the pursuer trotted "deliberately" toward the pursued male with head low, ears back, and horns flattened until the latter either left or was chased from the area. The pursuer followed for a few meters and then pulled up abruptly in a head-up display with vigorous head-shaking.
>
> [When a female entered a male's territory] he typically ran up to and around the female, trying to prevent her escape without physical contact. If she left the territory, the male's chase ended abruptly in a display, thus allowing the next male to take over, and so on in relay until the female either chose to remain within a male's territory or left the lek. Males entering a lek were often chased successively in the same manner as if the lek occupants were an elite "corporate body" defending the lek as a unit.

When females did stop within the lek, Schuster notes, usually between ten and twenty would be tightly clustered around one or two males.

These three categories of male-male competition—scrambling for females one at a time, monopolizing resources that females will approach or areas they will travel through, and displaying to them in aggregations—often borrow elements from each other and, even as a whole, do not begin to exhaust the possibilities. There is a bug (*Xylocaris maculipennis*) which appears to be able to insert its own sperm in the sperm duct of other males, who then use it during copulation instead of their own. And then there is a parasitic worm (*Moniliformis dubius*), the males of which, after they mate, have the practise of sealing off the female's genital tract with a congealing secretion. This presumably prevents other males from copulating with the female and competing with the first male's sperm. Another possibility is that the secretion helps prevent the male's sperm from leaking out. There is nothing unusual about mating plugs; they are found in a great many insects, reptiles, and rodents. What seems unique about this worm is that some *males* have been found with their genitals sealed up. The investigators believe that given the specific conditions under which this worm mates, it might well have been adaptive for males to take each other out of the mating pool by walling off their competitor's genital regions.[1]

The list of devices through which males compete with each other is enormous. The males of some species, after mating with a female, paint her with repellent, or at least antiattractant, pheromones to discourage other males. In one fly species (*Johannseniella nitida*) the male wedges into the female's tract and plugs it with his body. His genitals lock in place. Eventually the female breaks off the parts of his body that are removable, but his genitalia remain. A common competitive strategy is a sort of pseudotransvestism. Females often have a characteristic color and/or behavior pattern that reduces the hostility of males by serving as a gender cue. The males of a number of species imitate this behavior or show these colors, subvert the defensive hostility of the dominant male, lurk about his territory, and sneak matings at an opportune moment. Sometimes males combine to compete. Some species of frogs call in duos, or trios, or even in quartets (within a larger lek display); each member of the group sings a different part. If one is removed the others stop singing and begin to search, apparently looking for another partner with the right range. This searching will stop if a new "partner" is reintroduced by playing a tape of the old partner's song. In a population of wild turkeys observed in southwestern Texas, two- or three-member bands of male turkeys (brothers) courted females synchronously, strutting, fantailing, and pinwheeling about her with the precision of a drill team. Juvenile males in some tropical-fish species, among them the common platyfish/guppy, will delay their maturation if kept in a tank with an adult male. One interpretation is that these males are able to use their maturation cycle as a competitive instrument. They stop growing when they reach maturity; larger males probably are more successful. Therefore, the theory runs, the juveniles deliberately retard their maturation so as to give them the best chance of growing larger and beating the neighborhood dominant.

Mortal battles between males are not common, though they do occur. A weaker male does not want to pick a fight he would be likely to lose; nor would a stronger male be selected to risk a wound that would make his victory pointless. A period of symbolic conflict, of "showing off," is therefore adaptive for both parties. These are "swagger matches," in which each male attempts to demonstrate to the other that his combative equipment, innate ferocity, agility, raw power, and endurance are all so overwhelming that no one but a fool would try to match them. One of the more common forms of symbolic conflict is moving air by roaring or bellowing,* but males also commonly blow themselves up to look as large as possible, violently attack objects in the immediate area, and feint and dart and charge and shove without getting mortally serious. All during these displays both antagonists are calculating their chances in a serious fight. Here is an example, a challenge brought by one musk-ox (in Alberta, Canada) against a "harem" owning bull:

> [The recording biologists were attracted to the scene by] numerous throaty, lionlike roars. . . . The fight itself was initiated by the challenger who charged the harem bull with lowered head from a distance of about ten meters. The harem bull stood his ground so that both animals sustained the brunt of the impact on the bosses of their horns. They maintained contact for a few seconds after the clash and pushed vigorously against each other. On the cessation of contact the harem bull swung his head rapidly from side to side as he backed up and in doing so apparently struck the challenger lightly on the sides of the face and neck with the tips of his horns. Both combatants then backed away from each other some ten meters while swinging their heads slowly from side to side, then turned parallel to one another and "strutted" in a stiff-legged gait. The harem bull then turned his back on the challenger and "thrashed" a clump

* Tacitus said of German war chants: "By the rendering of this they not only kindle their courage, but merely by listening to the sound they can forecast the issue of an approaching engagement. For they either terrify their foes or themselves become frightened, according to the character of the noise they make upon the battlefield. . . ."

> of willows with his horns. Thereupon, the challenger turned away from the harem bull, strutted a short distance, and suddenly broke into a run which precipitated a chase of about one kilometer.

The naturalists who watched this challenge, Paul Wilkinson and Christopher Shank, explain that the head-swinging is likely to be a reenactment of one of the basic tactics of musk-ox combat, which is stunning the opponent with a hard charge to the head, slipping by his defenses while he is dazed, and then stabbing up into his side. Obviously enough information was exchanged during the interaction described here for the challenger to decide against a serious attack. Sometimes the decision can go the other way, of course: Wilkinson and Shank estimate that between 5 and 10 percent of the musk-ox males in the herd they were watching died from direct battles during the 1973 rut.

Tim Clutton-Brock, of King's College Research Centre in Cambridge, was able to experiment directly with the phenomenon. He had been watching the swagger matches of red deer stags and discovered that the rate of roars (among other factors) was an excellent predictor of the outcome of the match. The stag that roared fastest carried the day. Clutton-Brock made three tapes of this roaring: one at two and a half roars per minute, a second at five per minute, and a third at ten per minute, which is much faster than any mortal stag could ever produce. (A typical natural rate is six roars per minute.) He then planted loudspeakers in the rutting area and played these three tapes when stags with females came by. The two-and-a-half-minute tape produced little response from the dominants. When the stag heard the five-minute tape he herded his hinds together and threatened the loudspeakers. And when he heard the ten-minute tape, "he leaped to his feet and looked extremely worried." Red stags, incidentally, have a special motivation to develop ways of resolving a challenge without entering into direct battle. Clutton-Brock has found that a secondary mating strategy for males is to lurk around a dominant's females and steal them when he is distracted by a fight. He calls this the "sneaky fucker strategy". (Clutton-Brock believes that 20 percent of the red stags are permanently injured anyway in fights over their four years of reproductive life.) This technique of assessing combat potential indirectly has even been traced back to the dinosaurs. The hadrosaurs had long, hollow crests, the chambers of which connected to the nasal passages. One plausible suggestion as to the purpose of these structures is that they were powerful resonators through which males thundered at each other, just the way elephant seals do today.

The whole thrust of the system is to work toward shorter male lifetimes. This is not just because of the misadventures of combat; a male's mating success rises and falls very rapidly as he reaches and passes his prime condition. A male that lived for a long time in a slightly less-than-prime condition would father very few offspring, because during any single season he would be kept from mating by the males who were then at their peak. A male who instead trades off long-run survival for a more intense mobilization of his energies over the short run, who throws his all into the fray with no thought for the morrow, will, in many common mating systems, be more successful and, Darwinism predicts, will come to define the gender. It is the winners that control the system; losers have no voice. The males that make these arrangements work for them will be the ones to mate; it matters not at all how much havoc any given mating system might have wrought among the losers. Nor will the females ever be selected to modify their mating behavior so as to make things easier for males, at least not unless the carnage rises to such a point that they have to wait an inconvenient length of time to be fertilized.

CHAPTER

FEMALE INFLUENCES ON MALE SOCIETY

IF FEMALES WERE RESPONSIBLE FOR THE ORIGIN OF MALES (by expropriating the reproductive cycle) and their further development (by picking just those males that offered the best service and the highest quality), it seems natural to ask whether they might also have designed the nature of male society. In other words, do females control the intensity of male competition, and if so, to what degree?

Certainly the instruments for that control lie in their hands. If the females bunch together during the breeding season, then an individual male can defend more mates than he could if he had to travel around to females dispersed throughout the habitat. So when females group together, for whatever reason, they make male competition more stressful, demanding, and violent by raising the stakes. If the females of a population clock their cycle in some regular, predictable way, then a dominant male will be able to use his time most efficiently, and this will increase the importance of being dominant in the first place.

Or if all the females in a group become receptive at once, then a dominant would have great difficulty in controlling the situation; by the time he was finished with his first female, the game would be over for the year.

Lions are an example of creatures that suppress male competition in this way. The lionesses of a pride all come into and go out of a short (two-to-three day) period of heat simultaneously. During this time each lioness requires enormous amounts of energy for each effective fertilization—often mating every fifteen minutes over this period. It has been written that the males ejaculate each time (though how evidence was gathered on this point I cannot imagine), yet most matings do not result in cubs. Each *productive* fertilization therefore takes a lot out of a lion. This feature of lioness's reproductive physiology is thought to have resulted from selection to lessen the amount of competition between the males of a pride. When a new set of male lions invades and takes over a pride, they kill the cubs gathered by the previous set of males. Females therefore ought to have an interest in reducing the rate of turnovers, and, since two lions can put up a better defense than one, of retaining two males that are cooperatively disposed toward each other. Competition between the males of a pride exists, but, compared with that of other male mammals, is weak. These long, exhausting copulations surely help toward this end. By the time a male has finished fertilizing the three or so lionesses that are his norm, he would probably be too exhausted to compete for the other three, even if the second male hadn't already fertilized them, or the females weren't already passing out of estrus.

Given this degree of potential control, it is interesting that females, by and large, do nothing to lessen male competition. The pups of elephant seal cows, for instance, are sometimes crushed under charging bulls. The seal females might be expected to defend their pups by spacing themselves out, so that the males have to devote some time and energy to travel, or they might cluster in groups of hundreds or thousands, in which case the competition would become too taxing for any bull, no matter how formidable. Instead they group (usually), in nice, defensible units of a few dozen. Why is it that they cluster in just those patterns that seem to drive male competition to its most intense level? (The answer is not that the males impose the organization. It used to be thought that the males herded the females, and it is true that they can be seen trying; but the females are both numerous and agile and in fact move relatively freely from "harem" to "harem". Males almost never succeed in imposing control over groups of females in nature; the famous example of the hamadryas baboon males is virtually unique.)

Female behavior in nature often seems to enhance male competitiveness rather than reduce it. Female garter snakes appear to insure that they will be mobbed by the largest number of males by spreading the dates when they emerge from hibernation over several days. Sand-bee females dig toward the sand surface and then stop. This induces males to dig down toward them, and during those excavations, the males are repeatedly challenged (at a rate of four challenges per average excavation). It is very common to see females appear to delay mating until a number of males have been gathered together first. H. T. Gier gives an example from the life of the American coyote:

> A coyote family has its inception in midwinter when a female comes into estrus and attracts one or more sexually active males. . . . It appears that all the reactive, unattached males within an attractive female's territory join the parade and follow the female for days. I have seen as many as seven males following one estrous female. . . . The males follow for as long as four weeks. . . . Copulation occurred when the female stopped, nuzzled a male, then positioned herself to that male and lifted her tail. . . . From observations in the field and on penned animals, I am forced to the conclusion that the female makes the choice as to which male will be allowed to mate.

Observations like these suggest that in some cases females might have some reason to foment an even more intense level of male competition than would have existed. The most direct evidence on this point is incitement, instances when females can be seen, or heard, actually stirring up one male against another. Examples of female incitement are fairly common, and sometimes their purpose is easy to understand. Female toads (*Bufo bufo*) seem to use incitement in order to acquire the male most likely to fertilize more of her eggs. Female toads have reasons for strongly preferring large males. They release their eggs into the environment unfertilized and depend on the males (who clasp themselves onto the female's back) to cover the eggs with sperm. Only a large male can both maintain his lock on the female's back (he grips her under the arms and around the chest) and be physically capable of releasing his sperm over all her eggs. The smaller the male in proportion to the female, the lower the percentage of fertilized eggs.

However, there are not that many large males (*Bufo* females are larger than males *), and large females often end up with smaller males getting a lock on them than they would like. Two Oxford zoologists, Davies and Halliday, upon looking more closely into the matter, found that while large females are often seen carrying small males before spawning, during the act itself large females and large males were found to be paired. The zoologists made some experiments that established that a large male can dislodge a small one from a

* It might be pointed out that males in nature are usually smaller than females. There are very few invertebrate species in which the males are not smaller than females (this decides the question of overall frequency on the spot). Species of fish in which the males are larger than the females probably number less than a third of the total. Larger-male species are a little more common among reptiles and amphibians. They might constitute a majority of the bird species and certainly do of the mammals, but even within the mammal class, larger females are not rare. A recent study [119] shows that larger-female species occur in 30 of the 122 families of mammal species, ranging from a family of bats to three families of whales, including the blue whale, which means the largest animal on earth is a female. Larger-female species are also common among the few species of monogamous mammals. One reason why males tend to be smaller, at least in the smaller animals, is that they become reproductively functional at a younger age. Sperm takes less time to build than eggs. The old observation about male spiders hatching first and leaving the egg case before the females might be explained best not as a defense against incest, but just because the males are ready to go earlier. Also, where there are disadvantages to being large (clumsiness, conspicuousness, bulk in birds), males are freer to avoid them because sperm is so much smaller than eggs.

female's back. They point out that large males have a distinctive call (the bigger the male the bigger his voice box and the more basso his call will be), and conclude:

> . . . male-male competition is the mechanism by which the change in pairing from prespawning to spawning comes about. . . . Because it is the female who is in control of locomotion of the pair (her legs are free for swimming) it seems likely that she will be able to influence whether male-male competition takes place or not. If she has a suitable mate she may swim away from other males in order to avoid interference. If she has been grabbed by an unsuitable male then we suggest that one strategy she could use would be to swim over to where there are other males and thus invite male-male competition.

Large males would prefer to mate with a large female if possible, because large females lay many more eggs.

As a rule the rationale of female incitement is seldom so obvious. Yet it is quite common. In many squirrels estrus is announced by the female with a screech or a rattle, or a pheromone discharge, which brings all the males in the area after her in a rush. "If she stops vocalizing for a while," writes one zoologist of the African bush squirrel, "a male might begin to move away, but as soon as the female calls again, he will turn around immediately and move toward her." [148] Similar behavior (of females inciting pursuit by a number of males) has been seen in white-tailed deer, bison, rats, a land crab, mountain sheep, Pacific bonito (when ready to spawn the female swims in an exaggerated, wobbling motion that attracts males), and a number of birds, including the pintail duck, winter wren, and sanderling (the females of which initiate aerial chases). Courtship among the shelducks and sheldgeese (water fowl) takes the form of the male's alternately directing sexual displays to his intended mate and threat displays in some other direction. Paul Johnsgard writes:

> Female shelducks and sheldgeese are exceedingly aggressive birds and usually *Incite* [Johnsgard's emphasis] their males to attack almost every animate object. If the male responds blindly

and returns from the fray thoroughly beaten, it is quite frequent that the seemingly implacable female will promptly reject her mate and begin to court the victor, regardless of what species it may be!

And finally, I would like to pass on this report, made by an amateur arachnidologist in West Virginia of the behavior of a local web-spinning spider (*Verrucosa arenata*):

> One male was found four centimeters from the retreat of the female. Another male crawled up the upper foundation thread toward the retreat, periodically giving a series of four to five jerks on the thread. The female, resting in the hub during this time, responded each time with four to five jerks. When the second male reached the vicinity of the retreat, the males exchanged jerks, approached each other, and combat occurred (by this time the female had moved to her retreat), resulting in the eviction of the first male.[99]

It *looks* as though the female recruited the second male to fight with the first.

The species in which female incitement is best known is the elephant seal. George Bartholomew, who wrote the classic study of these animals in the early 1950s, specifically noted the intimate connection between competition and copulation. The dominant males "rarely engage spontaneously in copulation," he wrote. "Instead their sexual activities usually developed as an aftermath of the efforts of one or more subordinate males to copulate with females . . . over which one of the harem masters maintained dominance. Copulation appeared to be used as an expression of dominance rather than an end in itself."

He illustrates with the following example:

> On the hauling ground at West Anchorage, at low tide in the middle of the afternoon, a small female not more than six and a half feet long left the other females and headed down the beach toward the sea. As usual the female's departure caused no response from the male dominant over the part of the herd in which she had been lying.

> When the small female had crawled about thirty feet, she passed one of the subordinate males; the male remained motionless until she had passed, then roused himself, turned, overtook her before she had gone another eight feet, pinned her to the ground with the weight of his body, and after not more than five seconds began to copulate with her.
>
> Copulation had been under way less than fifteen seconds when the male dominant over the part of the herd from which the female had departed discovered the pair's activities. This particular dominant male had so far during the day shown no sign of sexual interest in any of the many females available to him, but now, after rearing up and vocalizing once with great force, he moved at top speed toward the copulating pair.
>
> The subordinate male, apparently unaware that the dominant male's vocal challenge had been directed at him, did not perceive the dominant's approach until the latter was only five or six feet away. The realization that he was being charged so disturbed the copulating male that he instantly headed for the water as fast as he could go, without even bothering to disengage himself from the female.
>
> The female, whose back had been bent almost double by the subordinate male's hasty departure, did not move away after the male pulled free. As the dominant came abreast of her he stopped abruptly, paused a few seconds watching the flight of the displaced male, and then promptly pinned the female down and began to copulate with her. . . . The dominant male's sexual response appeared to be a direct result of the challenge to his social position by another male and only secondarily a response to the female.

During copulation, Bartholomew notes, the male has to wrestle the female about for a short time; during this period she often will bawl a loud "protest"—the word is Bartholomew's—and flip her pelvis and hind flippers rapidly and repeatedly from side to side in an arc. She may also flip sand at the male and nip him on the neck. He seems to struggle to keep her quiet, pinning her down with his great bulk (males weigh about three times more than females), clasping her with his flippers, biting her on the neck, and striking her with his open jaws.

After a few minutes of this the female "acquiesces"; sometimes even arching her lumbar region to make her vulva more accessible. Bartholomew's observations suggest that the female controls the passage of sperm into her body:

> . . . as soon as his penis is inserted the male becomes almost completely passive. He lies half on his side and half on his belly, eyes closed, pelvis thrust forward, and one flipper holding the female. For the most part the two animals lie almost motionless, sometimes with their bodies at a marked angle to each other. . . . During the last half of the copulation the female assumes the active role. Mild undulations pass slowly through the posterior part of her lower trunk, and she flexes and extends her hind flippers gently. Often she performs a series of slight dorso-ventral flexions with the posterior part of the body which slide the vagina back and forth over the penis. . . . Copulation lasts three to seven minutes and is almost invariably terminated by the male.

Recently these "protests" have been studied intensively by Cathleen Cox and Burney J. Le Boeuf. They found that males mounting estrous females are more than twice as likely to be driven away by other males when the female protests than when she remains silent. Thus the picture that emerges is that the females have a substantial degree of control over the fertilization process, and that they use this control, at least in part, to play males off against each other.

Why do they do this, if they do? What could be the point? Perhaps the single observation that can be made about the range of ways one male is chosen out of the many that are called is that physical exertion is almost always involved, or demanded. It might be that the female is interposing a test that prevents her mating with gross misfits, from pairing her genes with genes that are defective.

The case that fits this suggestion best is male competition among those creatures in which the males have only one gene set instead of the usual two. A male of such a species has only one source of genetic information. If one of his genes is defective he must rely on it, use it, and express it, regardless. Mating among the social insects—a prime example of these one-gene-set males—is just as strenuous as among the larger-bodied species we have been reviewing. In some cases a queen flies to great heights so that only the fastest, highest-flying males can reach her; on others she might crawl along the ground, requiring the males to search for her over periods of as long as several weeks.

Another kind of "test" is more familiar, relying on leklike swarms of males all actively battling each other for access to females. An example might be the mating behavior of a seed-eating harvester ant (*Pogonomyrmex*) that lives in the Arizona desert. In at least some sections of the desert there are six *Pogonomyrmex* species living together. Each species has its own mating site and its own time slot for visiting that site. (How all the nests of the same species in an area get to know which site is theirs, how to get to it, and when to visit, is an interesting mystery. Bert Hölldobler of Harvard has studied this ant and reports that the sites in his study were flat stretches of ground for two species and bushes and trees for two others. To the entomologist the chosen sites seemed indistinguishable from numberless others in the area, but, generation after generation, each species' reproductives congregate in the same spot, flying in from nests scattered all around the countryside.) The first males to arrive release a sweet odor, detectable to humans, which attracts both the females and more males. While waiting for the females to arrive these early males "run about in a frenzied manner." As soon as each female alights she is immediately surrounded by three to ten males. "At the height of the activity," Hölldobler writes, "thousands of such mating clusters carpeted the ground. As many as fifty clusters could be found in one square meter."

The first male to reach the female grasps her thorax with his mandibles and forelegs and tries to insert his organs. The female twists her abdomen away from the probing organs, denying them entrance. Other males grasp the first male and pull at him, trying to tear him off the female's back, which they sometimes succeed in doing. The first male reacts by clinging tightly to the female, so tightly, in fact, that the female might be cut in two. Eventually the female allows the male to mate with her. After a few minutes he relaxes his grip on her thorax and, while still hanging on, and with his organs still inserted, rolls over to her lower side where he massages her gaster (abdomen). (This might have something to do with sperm transport.) A second male then catches hold of the female's thorax and, while still more males tug at him, waits for the first to uncouple. Then the female begins to bite the gaster of the first male, who decouples, often with such speed that his organs are left clinging to the female's genitalia. The second male then copulates, if he can, and is often followed by a third. Hölldobler showed that the females release an incitement pheromone from their poison glands which, if extracted and dropped into a cage of males with no females present, will induce homosexual mounting.

Thus whatever else mating activity accomplishes in these ants, it will surely exclude males with muscular or mechanical deficiencies. Perhaps the practise of making males from unfertilized eggs, and therefore with only one gene set, evolved in the first place because of the benefits of giving a stress test to the genes being passed on. (The model works best if we assume a context of brother-sister matings; perhaps in isolated colonies.) In any case making males from unfertilized eggs is fairly popular among the invertebrates; it has arisen five times in the insects, once in the mites, and once in the rotifers. Sperm also have only one gene set, of course, and something like the present argument has been made about those sperm that fail during the race to the egg; that the egg, in effect, picks the sperm that suits it best. One researcher with a light touch was able to retrieve sperm that had ascended a female rabbit's genital tract; it was mixed with fresh sperm and reinserted into the uterus of a second female. Several aspects of the experimental design were thought to give an advantage to the fresh sperm, yet the sperm that had already survived in the first rabbit was a far more effective fertilizer. In other words, the ability of the sperm to endure the murderous world of the uterus (murderous to a sperm: 99.999+ percent of them die there) also says something about its ability to perform its fertilizing function competently.

One problem with the "stress-test" idea is that it only works in any clear-cut way with those genes important to surviving the test. The mates of the *Pogonomyrmex* females will be physically adequate, but how about mentally? Harvester ants have to be good foragers; where does that get tested? Nonetheless, all things considered, while the stress-test idea does not answer every question that idle intellectuals might dream up, it seems a sober and useful contribution. To state this hypothesis in summary form: A certain fraction of the male population (say 5 percent to 10 percent) is assumed to be genetically defective. The females that drift into using their tools of aggregation, asynchrony, and sperm control to foster male competition mate less often with defectives, and prosper thereby. They need take no other action, no evaluating, assaying, or sampling. Once they've set up the basic conditions the males will do the rest.

Still, there are other possibilities. Typically victory in these male-male encounters, including those incited by females, goes to the larger male. This might be an individual who has survived longer, or been more efficient at exploiting the territory, or both. If any part of that superiority is inheritable, then females might be selected to see that they mate with the largest male around. This might be called the *adaptive elite* theory. The suggestion is that male competition is useful to females, and therefore worth inciting, because it screens out the majority of merely adequate males in favor of a handful that are qualitatively superior in some way.

Each male constitutes an experiment in which a different set of genes is tested against the environment. A large male means that the experiment was a success; his genes are therefore good ones for females to mix with their own. This idea has been used to explain the peculiar practise that many species of tropical lizards (genus: *Anolis*) have of copulating in highly conspicuous places. To give a specific example, when an *A.*

garmani female is receptive, she perches on the exposed, lower portion of a tree trunk, faces down toward the ground, and elevates her rear slightly. If a male approaches her and is accepted, the two enter into a copulation that can extend for as long as twenty-five minutes. The elite theory—applied in this case by Robert Trivers—is that this behavior allows the females to bring three kinds of males into competition: large, dominant territory-holders; transient vagabonds; and small subordinate males that live surreptitiously at the borders of the dominants' territories. Of forty-nine matings observed, the score was: dominants, forty-four; vagabonds, three; and small subordinates, two. He believes that the females' public mating habits are responsible for this lopsided score, since by exhibiting themselves as they do they make it more likely that the dominant male, who is most likely to win any dominance encounter, will see them. If females profit from mating with large males, because these males have proved themselves best adapted to the conditions at hand, then this sort of exhibitionism would have a sound evolutionary foundation.

Cathleen Cox and Burney J. Le Boeuf have also applied this theory, in this case, to elephant seals. As noted earlier, they discovered that the protest calls that seal females voice during mating attempts act, whatever the motives of the participants, as incitement calls, since they double the chance that a second male will intervene and interrupt the attempt. Two patterns emerged from a study of these calls. First, females protested virtually all mounts at the beginning of their approximately week-long fertile period. (Females give birth to a single pup about six days after they arrive; the pup is nursed for twenty-eight days and then weaned, at which point the female returns to the sea. The females mate over the last few days before they return.) But as the day of departure nears, they protest less and less; by the last day a majority of mount attempts are not protested at all. Second, the mating attempts of young, small males were protested more often and more vigorously. "When young, low-ranking males mounted a female," the authors write, "it was apparent from their shifty eyes and nervous behavior that they were constantly monitoring the movements of nearby, larger males and were afraid of being attacked." At the very least these patterns show—if there were any doubts about it—that females can and do regulate their calls to some end. In the beginning of the estrus cycle, the top-ranking bull or bulls account for 90 percent or more of the matings.

> The manner in which the females incite males to compete for them must be clarified [Cox and Le Boeuf write]. One cannot distinguish whether females are simply protesting a copulation or attempting to incite noncopulating males to intervene and chase off a mounting male. From the observer's blind, it looks as if protesting females simply do not want to copulate, and this may be the case. The important thing is . . . the effect of her behavior on nearby males. A blatant, squawking, wriggling, sand-flipping female with a prospective suitor on top of her struggling to pin her down attracts the attention of all males in the vicinity . . . [and] sets in motion a sequence of male movements. For example a male dominant to the mounter may issue a vocal threat sufficient to move the mounter from the female. But before the aggressor can reach the female he is threatened and displaced by another male dominant to him. Several aggressive interactions involving other males may ensue, resulting in a considerable change in the spatial relationships of males in the area. Episodes like this usually end when the most dominant male in the area gets close to the female and prevents all other males from mounting her or he mounts her himself. . . .
>
> If females did not protest copulations, males would still compete for females and interfere with each other's copulations. . . . The effect of a female's protesting behavior is to intensify male-male competition and augment its consequences. Her behavior activates the social hierarchy; it literally wakes up sleeping males and prompts them to live up to their social positions. The result is that it is more difficult for young males to mate, and the breeding monopoly of a few adult males is increased.

Cox and Le Boeuf believe that old, mature males might be

especially attractive mates because mortality among males is high (about half the males of breeding age seem to die from one year to the next), and males that survive demonstrate valuable qualities by that fact alone.

The elephant seal females apparently can manipulate the system in the opposite direction as well, toward the relaxation of male competition. Most females—even those that copulated repeatedly earlier—mate at the very end of their season, just as they leave the rookery and begin to swim out to sea. The reason they do this, Cox and Le Boeuf believe, is that the few males that monopolize breeding during the major part of the season copulate continually, both day and night, sometimes for hour after hour after hour. Male mammals need high concentrations of sperm to fertilize effectively; a human with "only" five million sperm per milliliter of ejaculation has a very poor chance of paternity. It is possible that not even so formidable a creature as an alpha (top-ranking) bull seal can produce sperm fast enough to satisfy the demands being made on that supply. Thus a female that had copulated, even repeatedly, with an alpha male might not have been successfully fertilized. The last-moment matings can be understood as a fertilization insurance—preferably "bought" from a male who has not copulated at these high rates in the near past. The females therefore leave the harem bulls' sphere of influence and mate with a peripheral male.

> These males adopted the strategy of pursuing departing females. The male that copulated was usually a low-ranking male relative to males in the harem and one who copulated infrequently, less than ten times during the entire breeding season. . . . Only one out of every ten mounts directed to departing females was protested. . . . Whereas alpha males were usually the first male with whom a female copulated, in only three out of twenty cases was he the last.

This seems to show that the females can control their relationships with males through their control of the structure of male competition.

The adaptive elite theory crops up again and again. H. T. Gier, whose observations on the coyote were quoted earlier, said that in the mating system of that species (which, the reader may recall, involved a female's leading a file of males around behind her for as much as a month before she made a choice) . . .

> . . . Ovulation occurs two or three days before the end of receptivity, so in most cases, elimination of the suitors, either by discouragement by stronger males, dissipation of stamina, or rejection by the bitch, has been effective in limiting the sire of the pups to the strongest, most cunning male available.

Perhaps the most extreme formulation of the theory is the suggestion of Amotz Zahavi of Tel Aviv University that there exists a "handicap principle" in male evolution. He points out that it is common to see males undertaking what appear to be pointlessly grandiose adventures—the defense of a territory far larger than any they might need for strictly bioenergetic purposes, or (more subtly) the flaunting of bright colors or noisemaking in habitats where predators stalk. His suggestion is that females might have evolved to pick males who have successfully proved their quality in a "swim-the-widest-river, climb-the-highest-mountain" kind of test.

The handicap principle carries the idea of female leverage to an extreme. It is one thing to force the males to compete among themselves for access to females; no new adaptations are needed to imagine that process beginning. The handicap principle requires that a separate adaptation arise, the "handicap," and that it necessarily involve some of its bearers throwing their lives away. All one can say is that the rewards to the survivors would have to be enormous; females would have to be interested in no other issues (like the conveniences of where and when, or courting skill). And they would have to be predisposed to begin rewarding such reckless behavior as soon as it appeared, else surely so costly a trait would be abandoned, even by males, very soon. Yet, who knows.

There are variants of the genetic-elite theory that stress

specific competencies rather than ecological adaptiveness or general physiological superiority. Robert Trivers has speculated that the horns of male reindeer had been shaped so as to advertise an ability of metabolizing calcium. These horns, he points out, are not very effective as weapons. The sharp, knifelike horns of the mountain goat or the fistlike horns of the mountain sheep both seem better suited to that end. Reindeer males occasionally get their horns so tangled up that the combatants starve to death, which does not suggest good weapon design. Further, they seem most imposing when viewed from the side; head-on their points and branches tend to fall into the same planes and vanish. One would think that if these horns had something to do with weaponry, they would have been selected to look most impressive from the viewpoint of the opposing male, or head-on. Nor are they used against predators; in fact they are shed just before those seasons, fall and winter, in which predator attacks must be most severe. They must be expensive to produce, and certainly add enormously to the burdens of running (especially through low-hanging foliage!). Trivers notes that the horns of these species become larger, both absolutely and as a proportion of body weight, as we look at larger and larger species, up to and including the elk, caribou, and moose. The young of all these animals are able to run with their mothers shortly after birth (often in only a few hours). Thus they cannot be the size of cubs or pups when born; they have to be large enough to keep up with a running deer, and must also have a developed capacity to support their body weight over the range of demands imposed by hard running. Skill at extracting and processing calcium is therefore important to females since their daughters will have to be able to bring the bones of their embryos to a high state of development; the larger the species, the more intense this need will be, and therefore, the more important it is for males to establish their competence on the matter.

There is a third approach to this issue of female choice, incitement, and male competition. It is possible that females choose males purely because they are attractive, are socially adapted, and that incitement is useful just because it allows a comparative examination of the candidates, like a line-up. In other words, perhaps in some species most or all of the males around will, if given a chance, father offspring that will be roughly equivalent in their physiological quality and environmental competence. Where these offspring will differ will be in one point only: the sons of some of the males will be more attractive to females than the sons of other males. By this theory females will pick good-looking males for mates because if they do so, their sons will be more likely to be good-looking, and therefore mate more often themselves.

The reasoning goes as follows: Imagine a condition in which the females of a species are generally surrounded by a number of importunate candidates, all of whom are approximately equivalent in paternal value and with whom the females mate at random. Then assume that one female develops a preference for males with long tail feathers and only mates with the "prettiest" member, by this standard, of any group of candidates. The development of this preference neither adds to or subtracts from her reproductive success and she enjoys whatever the norm is for the females of her species, generation, and region. In the second generation her daughters will inherit this preference and her sons will inherit longer tail feathers. Her daughters will reproduce at the local norm, but her sons will do somewhat better: They will get both their normal allotment of matings from the randomly mating females plus they will have a competitive edge in attracting their sisters. Assuming that inbreeding problems are either not severe, or, because the population is small and isolated, an equal trial for everyone, the males with long tail feathers will leave more offspring than the ordinary males. The attractive males have also been carrying within themselves the genes predisposing females toward long feathers—though unexpressed, of course, because they are males. These genes now begin to show up in the daughters of the third generation. The proportion of females in the population that prefer males with long tail feathers will also begin to grow, riding the success of their fathers. When these females mate they will pick according to their taste, and when they give birth the exercise of that taste will have increased the number of attractive males, who in the fourth generation will further spread among females the genes for this mating criterion, and so on. The system reciprocally accelerates itself, and, on paper and in computers anyway, seems to show that if there is no better idea around, females will rapidly be selected to winnow down their choices among the excess males by the assertion of aesthetic criteria alone. Peacocks come to mind as an example of a species in which it looks as though the males have become objects of art, pleasant to contemplate, defined by and selected for aesthetic terms and functions alone, and having no economic function. (But it is also possible that long tail feathers do communicate something about the physiological or ecological adaptiveness of the male. They certainly fit Zahavi's handicap theory aptly enough.) Birds that display in aggregations, or leks, often appear to be competing on aesthetic grounds. One naturalist wrote about the displays of a European shore bird, the ruff *Philomachus pugnax*:

> Males with relatively high rates of copulation differ from other males on the lek with respect to the development of their nuptial plumage. On all leks observed, preferred males were characterized by an optimal development of the nuptial plumage. . . . Thus it is likely that size and brilliance of the nuptial plumage directly influence a female's choice. This conclusion is also supported by the observations of Selous who noted that "handsome" males characterized by a full plumage were selected most frequently by the females.[73]

CHAPTER

XII

THE COST OF MALE COMPETITION TO FEMALES

ANY POWERFUL TOOL CAN BREAK DOWN OR GO OUT OF CONTROL from time to time, and sometimes females find themselves being drawn to their cost into the fierce dynamics of male competition. In some cases their physical survival might be threatened, as in the case of the observations Merle Jacobs made on dragonflies (*Plathemis lydia*):

> In many cases the female is chased by other males or is seized while ovipositing [egg-laying]. . . . Many males may be involved in this process at one site leading to a melee in which they dash at one another swiftly. . . . In these cases the female, being unable to oviposit, may fly from the pond followed by many males, some of which have come from adjacent areas. Kingbirds may fly into these flocks and catch males, or, more frequently, the slower-flying females.
>
> When two males simultaneously begin seizure of a female she may be injured and drop into the water. Wolf spiders often catch such females.

Or the costs may be inflicted on the females' offspring. A famous example of this are the pups that are crushed by elephant seal bulls as they plow back and forth across the shoreline attacking and fleeing from each other. There is no reason why a bull should be selected to care for the pups of the males who fertilized the female last year, and they don't. A more insidious instance of how male-male competition can be destructive to a female's interests is the practise of infanticide, and this is a story worth developing at length.

The hanuman langur (*Presbytis entellus*) is a black-faced, gray-furred monkey that ranges all across the Indian subcontinent, from the Himalayas to the southern coastline. They are "about the size of a springer spaniel," Sarah Blaffer Hrdy, who studied these primates for five years, writes, "and are endowed with the slender-waisted elegance of a greyhound." Their society is organized in the standard mammalian pattern. A band of females accompanied (usually) by a single male. The females are related, remain in the same territory and society all their lives, and recruit new band members from among their daughters. The males compete among each other for residency rights; the right to live with the females. Sons are driven away from the troop when they mature and will only rejoin troop life when and if they succeed in usurping a resident male of some other troop.

The actual displacement is accomplished by a group of males who first chase the former resident away and then quarrel among themselves to see who gets to stay. Even males securely in residence in one troop of females will reconnoiter the situation in neighboring troops and attempt to take them over and succeed in being a resident in two troops at once. The result of all this restlessness is that sexual change is a way of life among the langurs; males come and males go. On average, a newly resident male can expect about twenty-seven months of tenure before another male defeats and ejects him, just as he had ejected his predecessor. (This is about the same length of time as it takes for one of his daughters to reach sexual maturity. A troop that changes its male every two and a half years thus prevents inbreeding.)

In 1963 some Japanese primatologists were tracking langurs through the teak forests of southern India when they saw a band of seven langur males drive out the "leader" of a troop.

One male among the seven remained with the females. Within days all six infants were bitten to death by the new male. This report seriously conflicted with the prevailing belief that a healthy, "natural" society is one that harmoniously enlists the energies of all its members and functions for the good of all. The observation by Sugiyama was interpreted as a symptom of some pathology or other, no doubt induced by the malign influence of human civilization.

Hrdy spent two to three months each year, from 1971 to 1975, watching five langur troops that lived on Mount Abu in Rajasthani. Among these troops were two she called Hillside, whose resident male was named Mug, and a neighboring troop called Bazaar (because its inhabitants foraged in the local bazaar).

> In June 1971 the Hillside troop contained one adult male, seven adult females, six infants, and one juvenile male. In August of that year, Mug was replaced by a new male, Shifty. . . . At the time of the takeover, one adult female and all six infants came into estrus and solicited the new male. [Copulations are initiated by the females.] Local inhabitants witnessed the killing of two infants by an adult male. Each killing took place at a site well within the range of the Hillside troop; in fact, one occurred at a location used extensively by that group. It seemed highly probable that the missing infants had been killed, and that the usurping male Shifty was the culprit.
>
> On my return to Abu in June 1972, I was surprised to find that the same male, Shifty, had now transferred to the neighboring Bazaar troop. In 1971, Bazaar troop had contained three adult males, ten subadult and adult females, five infants, and four juveniles. Three of these infants were now missing. The killing of one had been observed by a local amateur ornithologist who lived beside the bazaar. The three Bazaar troop males remained in the vicinity of their former troop; the second-ranking of these bore a deep wound in his right shoulder.
>
> During 1972, Mug took advantage of Shifty's absence to return to his former troop. At this time Hillside troop consisted of the same six adult females and their four new infants. Two females, an older, one-armed female called Pawless and a very

old female named Sol, had no infants. Although Mug was able to return to his troop for extended visits, whenever Shifty left Bazaar troop on reconnaissance to Hillside troop, Mug fled. On at least eight occasions, Mug left the troop abruptly just as the more dominant Shifty arrived, or else the "interloper" was actually chased by Shifty. Typically, Shifty's visits to Hillside troop were brief, but if one of the Hillside females was in estrus he might remain for as long as eight hours before returning to Bazaar troop.

During the periods Mug was able to spend with his former harem he made repeated attacks on the infants that had been born since his loss of control. On at least nine occasions in 1972, Mug actually assaulted the infants he was stalking. Each time one or both childless females intervened to thwart his attack. Despite their heroic intervention, on three occasions the infant was wounded. During this same period, other animals in the troop were never wounded by the male. When the same male, Mug, had been present in the troop in 1971, he had not attacked infants. Similarly, during Shifty's visits to the Hillside troop in 1971, his demeanor toward infants was aloof but never hostile. Whereas Hillside mothers were very restrictive with their infants when Mug was present, gathering them up and moving away whenever he approached, these same mothers were quite casual around Shifty. Infants could be seen clambering about and playing within inches of Shifty without their mothers taking notice.

In 1973 Mug was joined by a band of five males. Nevertheless, the double usurper Shifty could still chase out all six males whenever he visited the Hillside troop. A daughter born to Pawless during the period when both Shifty and Mug were vying for control of Hillside troop was assaulted on several occasions by the five newer invaders; the infant eventually disappeared. . . .

By 1974 Mug was once again in sole possession of the Hillside troop and holding his own against Shifty. When the Hillside and Bazaar troops met, Mug remained with his harem. On several occasions, the newly staunch Mug confronted Shifty and in one instance grappled with him briefly before retreating behind females in the Hillside troop. Mug resolutely chased away members of a male band who attempted to enter his troop. . . .

When I returned to Abu in March of 1975 Shifty was no

longer with the Bazaar troop. In his place was Mug. . . .

Mug's former position in the Hillside troop was filled by a young adult male called Righty Ear. Righty (with a missing half-moon out of his right ear) was one of the five males who had joined Mug in the Hillside troop two years previously. Since that time, Righty had passed in and out of the troop's range, traveling with other males but not (so far as I knew) attempting to enter the troop. Righty's "waiting game" apparently paid off in March, when he came into sole possession of the Hillside troop. But, as in the case of his predecessors, Hillside troop was only a stepping stone: In April 1975, Righty replaced Mug as the leader of Bazaar troop.

The first indication I had of Righty's arrival in Bazaar troop was a report from local inhabitants that an adult male langur had killed an infant. On the following day when I investigated this report . . . an elderly langur mother still carried about the mauled corpse of her infant; by the following day she had abandoned it. Righty subsequently made more than fifty different assaults on mothers carrying infants. Nevertheless, only one other infant disappeared. Five infants in the Bazaar troop remained unharmed when my observations ended on June 20.

After Righty switched from Hillside to Bazaar troop, there followed some nine or more weeks during which the Hillside females had no resident male except for brief visits from Righty. Whenever the two troops met at their common border, Hillside females sought out Righty Ear and lingered beside him. These females were fiercely rebuffed by resident females in the Bazaar troop. Hostility of Bazaar troop females toward "trespassers" from Righty's previous harem prevented a merger of the two. The troops were still separate . . . in October 1975, but the vacuum in Hillside troop had been filled by a new male, christened Slash-neck for the deep gash in his neck.[79]

The sad story of the Hillside troop is obviously exceptional; otherwise there would be no langurs (during this period infant mortality at Hillside reached 83 percent). Another troop in the area, presumably more normal, retained the same male from 1971 to 1974. But while the example is extreme, the phenomenon seems to be a general one.

Hrdy cites data gathered from three different locations—

fifteen takeovers in all. At least nine coincided with attacks on or with the disappearance of unweaned infants; thirty-nine infants are known to have vanished during takeovers. These attacks only occur when the males enter the troop from outside, generally for the first time, and in any event, as in the case of Mug's first being ejected by and then replacing Shifty, after enough time has been spent away from the troop for infants sired by another male to be born.

The key to this story is an adaptation langur females have evolved to the loss of their infants (perhaps originally through attacks by leopards). Ordinarily they have a single birth every two years. But if they lose an unweaned infant the females immediately return to estrus and become receptive again. Seventy percent of the females in Hrdy's sample who lost their infants gave birth again within six to eight months. Thus a male that kills the infants sired by his predecessor substantially advances the time at which females in his troop will begin bearing his offspring. Infanticide gives him a competitive advantage over other, more tolerant males. To put it another way, a gene disposing a male toward infanticide would spread rapidly in a mating system like the langurs'. Further, the more intense male competition is, and the higher the rate of turnovers, the more important such a strategy will be. Infanticide on this pattern has been found in lions and in every major group of primates.

Male-male competition can work against the interests of females in other ways. Certain mating interactions in some species (the orangutang; some ducks, such as the mallard and the green-winged teal) look remarkably like rape, though rape seems strikingly rare in the animal kingdom as a whole. There are some birds in which the males are monogamous only so long as opportunity, in the form of other females, does not present itself, at which point they will spread their parental energies over two or three broods. This is presumably good for them but places an added burden on the females, or so one must assume. Theoretically, male-male competition would select males that speed through a series of copulations fastest, even if that involves misleading, deluding, and outright lying to the females involved.

Clear-cut cases of male lying are very hard to find, perhaps because the females are too smart for them. Male deception has been invoked to explain the evolution of empty "nuptial gifts" among a group of insects called the balloon flies. These are flies that have been found to court females by offering them little silk balloons. A leading authority on these flies, Edward Kessel, believes that the practise took its first evolutionary step in a species in which "the male avoids any cannibalistic attention on the part of the female by carrying with him, ready for presentation at the moment they embrace, a wedding present in the form of a juicy insect." These males then evolved the habit of using silken threads to quiet the captured insects. Next the silk threads became more elaborate and began to share, with the still relatively large prey, the function of stimulating and attracting the females. The balloon appears in this stage. They are composed of beadlike chains of tiny bubbles, wound spirally into thin-walled, white or transparent, iridescent spheres that look like tiny soap bubbles. As the bubbles became more attractive, the balloon took on a decorative role and eventually the prey vanished entirely. Finally, in Kessel's last stage, the male spins his balloon without incorporating any elements from prey insects at all.

The theory behind this is that the females were so taken with the little silk baubles that they no longer demanded real food. Assuming that the males are able to reuse the balloons over and over again, then spinning one balloon once rather than catching a number of insects might yield a competitive advantage. Still, deep down inside I cling to my prejudice that the balloon-fly females are not being fooled, and that the last word on these species has yet to be written.

Is there anything females can do to control the ravages that male-male competition can wreak upon them? One possibility is that they might just make fewer males. Theoretically this might be managed. One can imagine a grand council of all the langur, or dragonfly, females from an area meeting and agreeing that the services that males provide could be met by a ratio of one son to every ten daughters. No doubt, continuing with this fantasy, all the females would agree that the idea had great appeal, but whose son would be selected? Obviously every male in this "reformed" system would father ten times the number of offspring as each daughter would mother. Unless the females imposed a convention on themselves, reinforced by policing and sanctioning mechanisms, cheating by making sons would be automatically selected. (Here we see the same "anarchistic bent" that prevents predators from establishing quotas or males from organizing themselves into a common front.)

So far as is known all females that live in fully interbreeding populations divide their energies equally, on average, between the production of sons and daughters. Some parasites do not live in interbreeding populations; brothers fertilize sisters, and sons fertilize daughters. In these cases the mother can constitute herself as a government of one and does indeed lay just one male to a brood of daughters. The social insects do put less effort into making males than females, but in these species the male role as a genetic emissary of the nest is reduced. If one corrects for this by looking at the ratios of energy invested in each sex in proportion to its genetic significance, one finds, as Trivers and Hare did, that these ratios still equal each other.

This homeostatic equilibrium does have a loophole, since it does not apply strictly to the production of equal numbers but to the mother's energies, to how she trades off investment in each sex. Sons and daughters should each take about the same size bite out of her reproductive life. If sons take more energy to make, perhaps because they are bigger, then fewer of them will be made.

Suppose that something in the environment kills males more often than females. The sex ratio among adults would be unequal, and those males that survived would indeed father many more offspring than the average female. But mothers would still not be selected to make more sons, because so much would have to be taken away from making daughters to be confident of raising just one more reproductively effective, surviving son that it would just not be worth it. The bottom line for the langur females is that if leopards regulate the numbers of males, then the pressures of male-male competition will slacken off and infanticide with them. There can be no selection for females to make more males if some objective, implacable, unmanipulatable force is doing the regulating. But the langur females cannot do it themselves, because it is their own sons that are the problem.

In fact, in certain circumstances, a mother might actually be selected to allow or even promote the exploitation of her daughters as an instrument in male-male competition. This would happen if the benefits that flowed to her sons were large enough to outweigh the costs levied against her daughters; such a female might end up with a net advantage in grandchildren over one that did not participate in that behavioral system. For instance, the langur males that first evolved an inheritable tendency toward infanticide might well have had an extraordinary reproductive success, if langur society was anything like it is today, organized by residents, with male invasions and turnovers. If so, females that passed on to their sons a tendency toward tolerance would have found themselves selected against.

But this could only work over the short run, while the genetic context for infanticide was spreading and there were "tolerant" males to compete with. Eventually all the males would be infanticidal. The average reproductive success of males carrying the genes predisposing toward infanticide would have returned to that which males had had before the behavior evolved at all. There would be no one left to compete with (on this issue). What would be left would be all those slaughtered infants; all that wasted female energy. At this point, if and when a counter-infanticidal strategy emerges, it could spread and succeed on its ability to bring to maturity those infants that would have been killed.

At least that is the theory; unfortunately for the theory it looks as though infanticide is both universal and very old in this whole subfamily of monkeys, the colobines, or leaf-eaters. It is easy to see why, in this case, males cannot just simply stop killing infants. By the time a tolerant male had come to the end of his tenure as resident (twenty-seven months on average), most of the infants which he had patiently waited to father would not yet even be weaned, and would be murdered by the next, presumably infanticidal, male.

Hrdy did see a number of signs of female counter-strategies. When a new male takes over, certain females of the troop, especially the old and childless, may combine forces to resist his attacks on the troop's infants. Pregnant females sometimes display solicitation signals to a new male; since they could not possibly be ovulating at the time, this might be an attempt to confuse the male about the paternity of the infant-to-be. Once a female in a recently usurped troop, who had been traveling apart from the new troop (perhaps to avoid the new male's assaults), was seen leaving her partially weaned infant in the company of another mother. She then returned to the main body of the troop alone. In three instances females with unweaned infants left recently usurped troops to spend time in the vicinity of males that Hrdy believes were the fathers of their offspring. When a strange male or group of males invade a troop, the females fight with the current resident against the invaders. Usually these struggles involve cuffing and chasing and noise-making, but sometimes they go beyond that. In 1882 a Victorian naturalist watched an especially bloody encounter:

> . . . Two opposing troops [were] engaged in demonstrations of an unfriendly character. Two males of one troop [apparently these were the invading males] . . . and one of another—a splendid-looking fellow of stalwart proportions—were walking around and displaying their teeth. . . . It was some time—at least a quarter of an hour—before actual hostilities took place, when, having got within striking distance, the two monkeys made a rush at their adversary. I saw their arms and teeth going viciously, and then the throat of one of the aggressors was ripped right open and he lay dying.
>
> He had done some damage however, before going under, hav-

> ing wounded his opponent in the shoulder. . . . I fancy the tide of victory would have been in [the other invader's favor] had the odds against him not been reinforced by the advance of two females. . . . Each flung herself upon him, and though he fought his enemies gallantly, one of the females succeeded in seizing him in the most sacred portion of his person, and depriving him of his most essential appendages. This stayed all power of defense, and the poor fellow hurried to the shelter of a tree where leaning against the trunk, he moaned occasionally, hung his head, and gave every sign that his course was nearly run. . . . Before the morning he was dead.[79]

As diverse as all these strategies are, they are not effective. The infants still die, and, as we saw in the case of the Hillside troop, sometimes at such great rates as virtually to ensure the extinction of a troop. It is something of a mystery why the langurs have not succeeded in breaking out of the box into which these excesses of male-male competition have led them. Hrdy mentions two possible counter-strategies. The first is that the females could organize. They could refuse to copulate with an infanticidal male, or not advance their ovulation cycle after one of their infants died, or form a unified force that would punish infanticidal attempts severely. Or they might grow larger than the male; this would put single females in a position to punish infanticidal males (presently langur males weigh about eighteen kilograms; the average female weight is twelve kilograms). It has been proposed that the reason why the hyena female is larger than and dominant over the male is for just this reason, to prevent males from threatening her cubs. But for whatever reason or reasons the langur females have not taken either route. The point of this melancholy tale is that systems of male-male competition can arise that end by leaving both sexes trapped in a behavior that is to the advantage of neither.

CHAPTER

MALES WITH AN EDGE

A MALE READING ALL THIS MIGHT WONDER IF THERE IS ANY end to the demands that females can make upon males. Do any members of the gender ever manage to climb out of their inferior position? Do females always have everything their own way? Obviously to the degree that female demands drive down the number of competing males, either by killing them through exhaustion or by forcing them to spend a lot of time over each female, then females will have fewer males to play off against each other and will have to temper their levies against male energy. So there are some limits built into the system. Since male competition has, in some species, made males larger than females, males sometimes have a competitive advantage in nonmating interactions, like squabbling over food. (Though in actual practise "dominance" can shift back and forth, depending on the issue, the reproductive state of the female, and other factors besides size or weight.)

One exception to the picture of males conforming to female needs is the lek displays, in which males seem to organize themselves and pull the females toward them, rather than the other way around. Lek displays have been interpreted here as existing because they allow females to shop for mates among a larger number of candidates and pick the single best—however "best" might be defined—of all the males in the area.

Thus the real actors in the system are not all the males and all the females, but a tiny handful of highly attractive males—a large population of peripheral males whose unwilling role is to provide a background against which the winners can identify themselves to the females. Despite there being dozens or hundreds of physical males on the lek, there are usually only one, or two, or three reproducing males. These reproducing males have a much more powerful leverage position vis-à-vis the females; in effect they get the females to inconvenience themselves as much as the males, in that both sexes now have to travel to the lek. (If there were only one acceptable male in the population he wouldn't have to go anywhere; all the females would have to find and pursue him.) Both the winning males and the females exploit the peripheral males, who carry the burden of advertising the quality of the central males to the females, but get nothing out of it themselves except perhaps a learning experience. The greater advantage of central males can be seen in other ways than that they get the females to come to them. A study of lek displays in the greater prairie chicken (*Tympanuchus cupido pinnatus*) showed that central males spend less time courting females than do the peripheral males, and yet still get by far the preponderance of copulations.[11] If for some reason a preferred male is available only for a short period, then females might be selected to conform to his schedule rather than the one most convenient for them. Cox and Le Boeuf note that:

> Since there is a limit to how many females an alpha male can inseminate . . . it would be advantageous for a female to mate with him early, when he is fresh and before his fertility or sexual interest starts to decline. . . . We know that females who arrive early in the season form the center of the harem (and are therefore) more likely to copulate with the alpha male than females on the periphery, because the alpha male takes up a central position among the females. . . . Centrally located females are aggressive to females on the edge and keep them from entering the center of the harem, and early-arriving females may prevent late-arriving females from joining the harem.

The examples of male incitement given earlier in chapter eleven may have seemed like a tyrannical assertion of authority. Still, looked at from the point of view of the dominant male, male incitement is simply a request by the female that he step forward, brush aside some pathetic competition, and mate with her. What could be more convenient for him than that? Without her call he would have to keep all the males in the area continuously suppressed; incitement serves his interests by making it possible for him to arrange his time more efficiently. If female lizards (*Anolis carolinensis*) are exposed to the sight of male aggression, of males fighting over dominance position, their ovaries either will not grow or will shrink. They thus postpone their mating to conform to the dominant's schedule, until such time as he has vanquished the competition. They are dependent on the dominant for something, perhaps a quiet, trouble-free mating, or his genes, or both, and that dependence has forced them to pay attention to his problems. Female baboons have an estrus patch that flames out brightly when they are at the peak of their receptivity. Again, this is a great convenience for the dominant male, who can monitor the cycles without going to any more trouble than you or I might go to in glancing at a wristwatch.

There are a number of species in which females are dependent upon a resource which is controlled by males. The males that possess the richer resources usually get more females, which is an example of females indirectly accommodating themselves to males. Good examples of this are found in marsh birds that can live either monogamously or bigamously; whether a male lives with one female or two depends on the quality of his territory.

The African, or black-headed village weaverbirds (*Ploceus cucullatus*) have the potential for illustrating these points in a much more subtle and interesting way. The weavers belong to a group of twenty-seven species, all of which live in Africa, that build (weave) tightly woven, quasi-spherical, enclosed nests, usually with a downward-pointing entry hole or tube. The nests hang from the outermost twigs of trees, often over water. The point of these enclosed, hanging nests is thought to be protection from predators, including hawks, kites, and eagles from above, genets, wild cats, mongooses, and jackals from below, and a number of tree-climbing pythons. The village weaver gets the first part of its name from a taste for building near and in human settlements, and probably this is also, in part, an antipredator strategy.

In 1967 two California zoologists, Nicholas and Elsie Collias, spent seven weeks in the Senegal River Valley watching some colonies of weavers nesting in trees that overhung irrigation canals and streams. (The countryside is often arid and the weaver is dependent on these streams for drinking, bathing, the vegetation used to build their nests, and the insect life they feed their nestlings. Also, building directly over the water may be an antipredator defense.) The breeding season runs throughout the last two months of the three-month rainy season, which runs from July through September. One sign that breeding activities are about to commence is that the males change their plumage from the camouflage green-brown, with a yellow throat, to a stark yellow and black. Their bills change from gray or ivory to black; their heads from olive-green to black. Black patches appear on each side of their beak. The nape becomes a rich chestnut. The males then partition the branches of the breeding tree or trees among themselves with a range of aggressive interactions that range from simple and quiet exchanges to vicious head-pecking battles. The Colliases carefully distinguish among six different levels of aggressiveness, but it is clear from their detailed descriptions that they could just as easily have built a case for twenty. As an example, here is their description of level four:

> Song and full plumage displays are given during formalized border disputes or between two males seriously contesting for dominance. Sometimes a male holds a piece of nest material in his beak when defending his territory. . . . Sometimes one of the contesting males wipes his beak against a twig as if removing some distasteful object. The two males crouch forward facing each other with the neck drawn in and the head and body

feathers erected often to their fullest extent. The wings may or may not be quivered, depending on the intensity of the dispute, and the tail likewise may be fanned out only at higher intensity levels. As one male sings, the head is bowed, the beak is pointed downward . . . and the black head with blazing red eyes contrasts conspicuously with the yellow on body and wings. Frequently the birds sing alternately, each uttering some short preliminary notes terminating in a longer buzz. The song often seems to inhibit attack from the opponent for the moment, the birds appear to be listening one to the other, and may alternately advance and retreat, sometimes swaying the head slowly to one side as if with the intention to turn away and move off. . . .

After these intricate exchanges are complete the males begin, each in his own territory, to build the shells of breeding nests. The shells are woven from long (twelve inches to eighteen inches) strips of palm leaves, sedge, and grass. They are not crude affairs; for example, they have a double roof, perhaps, the Colliases speculate, as insulation against the sun and rain. "We have seen no instance in nature where the female ever helped the male to weave the outer shell of the nest," the Colliases write. "However, we have had one abnormal female in our aviaries who wove several nests by herself and even laid eggs in some of these after copulating with the male who held the adjacent territory." A male may build a nest in a single day. The "average" male built four nests in all. The males seem not to call continuously for females but wait for them to arrive. When one does:

. . . he goes to his most recently built, greenest nest and hangs upside down fluttering directly beneath the nest, showing the entrance to the female. . . . The inverted male clings with his feet to the inner lip of the entrance, flapping his widely opened wings and repeatedly uttering characteristic notes which we verbalized as *look-see! look-see!* . . . At the same time the male rocks or swings from side to side, all the while keeping his feet at the same place on the nest. . . .

The beating of the male's wings is made more conspicuous by the bright-yellow underwing coverts, and the spread wings reveal the yellow color on most of the feather shafts. . . . The

> whole effect of the continuous movement and shifting brightness pattern of the wings during the nest invitation ceremony is reminiscent of a flashing yellow light. A whole colony of males simultaneously displaying their nests to visiting females is a conspicuous and spectacular sight.

The female usually hesitates for some time in front of the displaying male, who may fly back and forth between her and his nest, singing constantly to her with a rich variety of songs. (The neighboring males also try to catch her attention.) If the female leaves the territory the male may build some fresh nests; if a specific nest has been repeatedly rejected, or begun to fade, it will be dismantled and a fresh one hung in its place. If a female enters the nest she will poke and pull at the materials for one to more than ten minutes, then may leave, return, and reinspect a nest several times before accepting it, if she does. (The hanging display of the male might be interpreted as being, in part, a stress test of the nest. Weavers breed in the rainy season, and their nests, which hang free, out in the open, can be violently tossed about by the strong winds and heavy rains.) If she accepts a nest the male begins to look around for another female (the "average" male has two) and she lines it, often beginning with a thin foundation of leaf strips, then a layer of soft grass heads or feathers. Frequently a male sings to a new mate each time she enters the nest with a beakful of lining material. He is excluded from the nest; should he try to enter the female utters a loud "protest" call and the male "invariably" desists and leaves. When pecked by one of his mates in his tree territory he ordinarily yields to her and does not peck back, though outside his territory, on the ground, he will shoulder his way ahead of the females if any food is presented to them. The Colliases report that in their aviaries, if a male has no nest in which to sleep, he does not oust a female from one of the nests he built but sleeps in the open.

All this seems a very straightforward picture of a rather common system in which the females are all equally highly desirable to males, and all males uniformly uninteresting to females. But I have only mentioned one side. At the same time as the Colliases were noticing most females hemming and hawing and delaying copulation, they were also noticing some females accepting a nest and copulating with a male within a few minutes of arriving on the scene. They even saw one female fly right by a hesitating female and shoot into the nest herself. When a male is courting a female he sometimes drives off the other females in his territory. The reason he does this, the Colliases believe, is that resident females will sometimes interfere with the courtship of a new female. Once they were watching a male displaying a nest to a visiting female that already had a female resident within. Suddenly the resident put out her head, grabbed the male by the feathers on the top of his head, and hung on for about twenty seconds while the male tried to free himself. Then she ducked back into her nest. The male flew off and did not display this nest again for at least thirty minutes, after which time the Colliases stopped watching. Why should the female have interposed her desires in this way? The point is that males sometimes "change their mind" about which female they prefer, will pull an already-resident female out of her nest, send her off, and display that nest to a second female! Presumably the female that grabbed the male was (successfully) preventing any chance that "her" male's enthusiasms might be won over by some other female for whom she would be evicted.

These are the clues of a very different balance of power, one in which a male or a class of males, has something to offer which a female finds useful enough so that she learns to grab it without delay when it is offered and fears competition for it from other females. There is nothing in the Colliases' report that allows one to decide what that quality or resource is, but there are a number of possibilities. One is nest location. Nests in the upper parts of trees were found to be less likely to be occupied by females than those hanging directly over the water. The reason for this is not hard to find. Pythons attack the weaver nests by crawling out over a lower branch and raising their body up to the level of their target. If only a small percentage of the males are able to acquire territories on the most secure branches, females might compete among themselves for residency rights in those nests, and the males that controlled them would be presented with a choice of mates. A second possibility is nest construction. Perhaps a fraction of the weaver males are much more highly skilled (maybe because they are older and more practised) at building nests. If this edge in workmanship has important consequences, then females might compete to be the mate of an older male. The third possibility is good genes. If the early territorial battles define a small class of winners, then those winners might have genetic qualities that make them highly attractive as mates. What I would guess is that as males get older they get better in everything. They win more territorial battles, their nests are more skillfully built and located, they become more adroit at courtship, and more aware of the dangers of the environment. The prediction is that evictions will be found to be conducted exclusively by older males, because older weaver males might be in the position that is occupied by females in most species: they attract a surplus of mates and so are free to indulge preferences.

Another way in which a fraction of the male gender might find itself in a more favorable power relation to the females than males usually enjoy might be as members of a "chronological elite". This might occur in species in which females strongly prefer to be mated within a very short time, too brief a period for males to be able to travel to more than one female (or even as many as one; the evolution of this system requires that some females have to accept fertilizations outside the optimum period and suffer reduced reproductivity thereby). One species in which this seems to have happened is the Chesapeake Bay blue crab, *Callinectes sapidus* Rathbun, the "beautiful tasty swimmer".

Crabs, of course, have an external skeleton (the word "crustacean" comes from the same root as "crust"), which they cast off about twenty times over their life cycle. They then expand their body size by about 25 percent each time and rebuild their skeleton out of elements filtered from the sea water and recycled from their old shells. The new shell does not harden enough to provide good protection for seventy-two hours, and usually, when crabs molt, they stay buried in the sand for at least a day or two. Females mate once, just after they have completed their last molt. Obviously a female that first cast her shell and then broadcast receptivity pheromones while buried in this condition would risk being found by predators as well as males, while a female that hid effectively enough to escape from one might be missed by the other. The blue crab has responded to these exigencies by evolving a very elaborate system of courtship and mating behavior that will allow successful fertilizations to take place even while the female is at her most vulnerable.

A typical fertilization sequence might run like this: First the female releases a pheromone. If a male is nearby and attracted he will approach the female and begin his display. In this he raises his body as high as he can and walks around her on tiptoe, opening and extending his claws, and lifting his flatter, rearward, swimming legs up over his body to form two waving, parenthesislike arcs. All these behaviors make the crab look much larger and more formidable than he really is (to a human, anyway). During his display the male snaps his body backward and kicks up flourishes of sand with his legs. Sometimes the male will repeat this whole routine over and over, while the female rocks from side to side and waves her bright red claw tips. Sooner or later one sex or the other takes the initiative and the female ends up tucked beneath the male. He then swims off, holding her under him, searching for eelgrass or some other hiding place. There is some evidence that the search for a suitable site can continue for as long as a week, though two days appears to be standard. When an old coffee can, or patch of grass is found, the male stands over the female and makes a cage with his legs. The female undergoes her last molt, which might take two or three hours. William Warner, whose *Beautiful Swimmers* is a natural-social-history of the Chesapeake crabs and the Chesapeake crabbers, describes the act of fertilization:

> When at last the female lies exhausted and glistening in her new skin, he allows her some moments to rest and swallow the water that is necessary to fill out her weakened stomach and muscle tissues. This done, the male gently helps the female turn herself about—she may well have gotten impossibly oriented in the final throes of ecdysis (molting)—until she is on her back face-to-face beneath him. . . . When the female crab is ready, she opens her newly shaped abdomen to expose two genital pores. Into these the male inserts two appendages. . . . When all is in place, the female so extends her abdomen that it folds around and over the male's back, thus . . . preventing any risk of coitus interruptus. They remain thus for from five to twelve hours.

For another two days or so, until her shell hardens, the female is cradle-carried by the male. "To think properly about the blue crab," Warner writes, "it is first necessary to assume that the species can and will perform anything in its life cycle at any time, dead of winter excepted." That warning in mind, one can say that in terms of crude averages, mating begins in early May, reaches a peak in late August and early September, and continues into October. It takes place high in the bay, a long distance from the ocean, where the female spawns. Once a female has been fertilized she travels down to the Atlantic. Females that have been fertilized early in the year can make it in one trip; other females overwinter partway down and finish their journey in the subsequent spring. A spongy egg case containing from seven hundred thousand to over two billion eggs is constructed over the two months after fertilization. When the female reaches the ocean she passes her eggs from her ovaries past the reservoirs where she keeps and maintains the sperm. The mass of fertilized eggs is attached to hairs on the outside of her abdomen, where she carries them until they hatch. If conditions are good the female can generate and fertilize a second batch. After two months of growth the hatchlings have begun to look like crabs, though as yet they are only one-tenth of an inch wide. They begin to migrate up out of the ocean and southern Bay into the rivers and upper Bay. They overwinter in Virginia waters, south of the Potomac, and resume moving up-Bay in the following spring and summer, during which time they grow to adulthood and mate. The females then return to the lower Bay and ocean to spawn; the males remain in the brackish, up-Bay waters and continue to seek females.

Blue-crab males devote a considerable amount of time to their mates, four days or more, during which they are out of the mating pool. Especially when the mating season is just beginning, in May, acceptable males will not be as ubiquitous as males usually are. Yet a May fertilization seems to be the optimum for females, giving them time to reach the ocean that summer and perhaps even generate two spawn. Females that mate in the summer have to overwinter in the Bay and then spawn the following spring, nine months after having been fertilized. Aside from the added mortality this implies, there is some evidence that such females do not have the option of spawning twice.

Females therefore need males most when the sex ratio is at its lowest, and they respond by seeking males out actively. A branch of the crabbing industry, called "Jimmy potting," has sprung up to exploit this rare business of male hunting. Male crabs, called Jimmies, are placed in a pot and submerged, or tethered, on a string. Then, when females that have been attracted by the males appear, they are caught. Warner quotes one crabber: ". . . I remember one Jimmy I had on the line, now he caught seven wives just as fast as he could get them. I just stayed there taking them. Then all of a sudden he wouldn't catch any more. I couldn't blame him none. . . ." Jimmy potting can only be practised at the very beginning of the mating season, in May and early June. Thereafter, one assumes, enough mated females have left the population so that males tend to be on hand when the time for the final molt arrives; the females do not need to look for them, or depart from their foraging routines to follow a male scent.

Blue-crab behavior is extremely flexible. Sometimes the male attracts the female; sometimes the other way around. Generally the male goes through an elaborate courtship display, mentioned above, in which he seems to be demonstrating his capacities for dealing with the responsibilities he is about to undertake—showing his height, and thus what kind of cage he can make, his breadth of reach, and his strength—but the female also (sometimes) waves her brightly colored red claw tips, an action that can plausibly be thought as intended to stimulate the male. (Only the female thus "paints her fingernails," to use the Bay expression.) At other times the female simply scoots under the male with hardly any preliminaries at all. Sometimes females are seen rejecting males, or mating with more than one. One possible explanation of this variability is that the blue crab is adapted to a breeding season in which the value of males begins high and then steadily sinks all summer, and that crabs have evolved ways of assessing this swinging balance of desirability and altering their mating behavior accordingly.

Over the last few pages of this chapter different means have been reviewed as to how a fraction of the male population might climb out of the inferior position in which most males find themselves vis-à-vis females. All the systems presented left the majority of the males, over a majority of the breeding season, as the "go-fer" of the mating interaction. The only way in which the majority of males in a species can have an equal relationship with their females is by attacking the root of the asymmetry between the two—the difference in reproductive tempi. In the next chapter we will look at some species that have managed to reduce this distinction between the genders.

CHAPTER XIV

MALE LEVERAGE AND EQUALITY

THE USUAL RELATIONSHIP BETWEEN THE SEXES CAN BE reversed for a number of reasons. First, if there are so few males that females cannot afford to wait for more than one candidate to appear, then females will lose their usual leverage. When Merle Jacobs, as part of his study on dragonflies, removed all but a few males from his study pond, he found that the dragonfly females took to pursuing and pouncing on the remaining males. The males copulated at abnormally high frequencies and finally left the pond altogether, presumably having found themselves in the unusual position of running out of sperm and energy before running out of mates. Obviously when courtship (or just the fact of being male) becomes so taxing that it removes males from the mating population, then females will have fewer males to play off against each other and will have to stop making demands on their time and energy.

Second, females might very strongly prefer, as mates, just a fraction of the available males. If a good male is hard to find, then assuming that the female cannot wait forever, she will have to go and find one herself. A male might possess an exceptionally attractive resource of some kind, one that cannot support all the females that would like to gain access to it. A propertied male of this sort might even be selected to pick and choose from among the females contending for access to his territory. The issue of females' being attractive or not would become important. This would be a 180 degree reversal of the usual mating relationship.

A third class is made up of cases in which males do not compete for more than one female. At an extreme this might be because copulation is fatal or severely debilitating to the male. Sexual cannibalism in mantids and spiders is one example. A second example, surely very unusual, has been found in an Australian marsupial shrew (*Antichinus stuarti*); copulation unleashes in the male a fatal surge of corticosteroids. The males of a group of neotropical frogs (genus *Atelopus*) fasten onto the backs of females weeks and even months before those females are ready to lay their eggs. For some reason the female does not scrape the "piggy-backing" male off and carries him until she needs her eggs fertilized; during all this time the male does not eat and grows progressively more emaciated. While little is known about this group of species, it seems unlikely that these males will be able to mate with more than one female in a season—maybe in their life.

Examples like these might be considered to be the simplest form of monogamy. Usually what we mean by that word involves a period of association between the genders, often, though not always, extending over more than one breeding cycle. The firmest of these associations are obviously those built on the model of the angler male; I know of no other in which the male is fed, like a fetus, via the female's blood stream. There are quite a number of parasitical and other filter-feeding invertebrates, such as some barnacles, that have adopted a "dwarf-male" system in which the sperm donor is either physically attached or lives in very close association with a much larger female. Monogamy often seems associated with female dispersion and a lack of mobility. If the females of a species do not herd or group themselves in any way, if they do not gather to compare mates at a lek, if they are not committed to a communal breeding site, then, given the rigors of travel, it may be impossible for a male to fertilize more than one female in a season. In these cases he is more likely to live in some kind of association with a single female and wait for her to become ready to breed over a number of seasons. For example, there are about two dozen species of small African antelopes, mostly forest-dwellers, that live monogamously in this way. Small antelopes (the smallest is only a foot tall at the shoulder) virtually never gather into herds. For one thing, their small size commits them to easily exhausted high-quality food items. (A small warm-blooded animal loses heat faster than a large one, which is why hummingbirds and shrews have to eat continually. A smaller antelope is therefore more likely to specialize on just the high-protein parts of plants.) Thus competition from other members of their herd would be especially burdensome; and their main defense is an ability to forage inconspicuously, which is hard to do in a group.

Both these life strategies require that the antelope have an intimate knowledge of every nook, cranny, bud, and twig in his/her territory. Members of these species thus tend to live solitary and rooted lives. The males and females may live as pairs or they may live in adjacent territories and associate only at mating time. In either case the costs of travel in these species are so high that males end up just mating with a single female.

Other circumstances that can force a monogamous society include highly unpredictable mating opportunities. There are two groups of parasitic birds that lay eggs in other birds' nests. In one group the males travel around competing for females with distinct, conspicuous displays. In the second, which includes the more familiar cuckoos and cowbirds, the two genders look alike and live monogamously. The difference seems to be that the monogamous parasites specialize on the nests of a single species, while the promiscuous birds (*Molothrus ater, M. bonariensis*) can lay their eggs in the nests of many host species. These generalist parasites lay more eggs (though perhaps their eggs are tossed out more often by the host parents, since their eggs are not such skillful copies of the hosts' as are the specialists'). Any nest parasite, whether generalist or specialist, has to lay her eggs expeditiously, during the period when the brooding bird(s) has left the nest to forage. If specialist parasites have fewer of these opportunities, which are not especially predictable, then they have to capitalize on them all the faster. Under these circumstances it may well be better for a male to attach himself permanently to a single female rather than risk visiting a number of females, but never arriving at quite the right moment. The more promiscuous generalists, by this theory, have enough targets so that the females can take the time to be courted.*

The fourth class of "powerful" males are those in which members of that gender provide some kind of parental care. This could include feeding and/or guarding the female; brooding, feeding, guarding, or tutoring the young; maintaining the site, or guarding the territory. All these forms are rare except in birds. Male parenting is virtually unknown among the insects—one-third of all animal species. One of the very few known exceptions occurs among the burying beetles, *Necrophorus*. E. O. Wilson gives a very succinct description of their habits in *Sociobiology*:

> In May the overwintered adults begin to search for the dead bodies of small vertebrates, such as birds, mice, and shrews. If a male encounters a corpse, he takes the "calling" posture, lifting the tip of his abdomen into the air and releasing a pheromone. The substance appears to attract only females belonging to the same species. If more than a single pair find a corpse—and sometimes as many as ten do—fighting ensues, male against male and female against female, until only a single pair is left. The winners

* Everything said in this section is based on the assumption that the adult sex ratio is about equal, and it may be that in some of the species mentioned here—the mate-eating insects, the anglerfish, the piggy-backing frogs—the females are in a vast minority, perhaps because they suffer differentially from the attentions of predators. In these cases the females that survive will have so much leverage that the males will have to fight and qualify for the opportunity to mate even once.

> then excavate the soil beneath or around the body, . . . chew and manipulate the putrefying mass until it is roughly spherical in shape . . . then seal off the burrow from below, entombing themselves with the rotting ball.

When the larvae hatch they sit in a crater formed by the female on the food ball. For much of their development they are partly dependent on regurgitated food; this is usually supplied by the female, though in two of the six European species of *Necrophorus* the male helps feed the young to a limited degree. (Two University of New Hampshire ecologists, Lorus and Margery Milne, studying American species of this insect, have noticed that in some cases, after the food has been buried, the male copulates and then buzzes off, leaving the female to minister to the young herself. But by then the most important piece of work, locating and secreting the food, has been accomplished.)

There are a few other examples of paternal care among the invertebrates. Some male crabs protect the female during her molt. Male fiddler crabs of several species dig burrows into which females enter to deposit their eggs. I have seen only one reference to male parenting among worms; and that was *Neanthes caudata*, a marine worm in which the male incubates the eggs.[122]

Paternal care is somewhat more common among the fish, though surely found in fewer than 5 percent of all species. Male seahorses and pipefish incubate eggs laid by the female in a special pouch. Paternal care is relatively widespread among the family *Cichlidae*, which are a very successful and species-rich group of fish. In some African species this care might involve mouthbreeding, in which one or both parents protect their eggs or young from predators by holding them in their mouths. In other species, often found in Latin America, the parents defend free-swimming broods, sometimes, from just-laid-eggs to "graduation," over periods as long as three months.

Kenneth McKaye, a biologist from Yale, has investigated

these fish and made some fabulous (and still very controversial) discoveries. One of these was his report that of the nine cichlid species that live in Lake Jiloa, Nicaragua, at least four adopt foreign young—young of other cichlid parents, sometimes even from different species—and integrate them into their own brood.

> In two instances parents were speared and the unguarded young were adopted by neighboring territorial pairs. We also saw territorial pairs take young of other pairs into their broods. This occurred twice when pairs were engaged in a fight with conspecifics. [Another] adoption occurred when one pair "kidnapped" some young of another pair. This happened when the fry of both pairs were approximately three weeks old, but the school of one pair was diffuse. When both parents faced away, the male of the second pair swam in and herded approximately 50 young out of the school and brought them back to his own school, which had about 150 young.
>
> In water of two to three meters in depth three pairs of *C. citrinellum* "communally" protected from predators a large school of four- to five-week-old fry. When danger threatened, the young would split into three groups and follow the pairs back into their respective caves. When the danger passed they emerged and mixed freely in the school. Individual fry were observed switching parents.

McKaye believes that if an adaptive explanation is appropriate for these phenomena, it is probably that up to a point a parent can decrease the risk that one of his or her progeny will be taken by a predator by increasing the number of foreign fry around his or her offspring. (This point would be reached when the cloud of fry grew so large that either the parents' defensive abilities were overwhelmed or it attracted a much higher rate of predator attacks.)

He found something even stranger. Some couples were helping their predators raise young! In eight cases he saw fish helping raise other fish that, when grown, might eat them. This might just be craziness, but, McKaye argues, not necessarily so. The altruist, *C. nicaraguense*, spends most of its life foraging out on the sand-silt bottom of the lake. It only approaches the lake's rock faces to look for holes in which to breed. The holes it needs are also used as breeding holes by other fish species, including one in particular, *Neetroplus nematopus*, that competes for the same size holes, at the same depths, and is a better fighter. These other species spend all their lives living near the rocks, as does the predator, who eats them. It is especially fond of *N. nematopus*. The predator has a hard time raising its young; all the prey species living in the rocks attack their fry, for obvious reasons.

Given these facts, McKaye argues that it might be adaptive for an altruist couple that has tried to breed but failed (because the competition for breeding holes was too intense) to raise a big crop of predators. (Their help makes a big difference to the success of the predator-parents, who do not bother the assisting couple in any way.) Then the altruist couple swims back out to the middle of the lake while "their" predators gobble up the competition that had been so bothersome. When they return they might indeed be subject to attacks from the very fish they helped raise, but they would also have access to breeding sites, and that tradeoff, McKaye suggests, might be worth making.

Paternal care in the amphibians and reptiles is also rare. There is one family of neotropical frogs (the *Dendrobatidae*) in which the eggs are laid on land and carried to water on the back of an adult. Sometimes this adult is a male. Kentwood Wells passes on this report of male parenting in the Panamanian poison-arrow frog (*Dendrobates auratus*), so called because a toxin secreted by its skin was used by the Panamanian natives to tip their arrows.

> I observed parental behavior performed by a captive male that mated with two females. After a clutch of seven eggs was laid, the male sat on them three times in ninety-five minutes. In each case, he sat on a wet rock in a shallow water dish for ten to fifteen minutes before sitting on the eggs. Twice he remained on the eggs for twenty-five to thirty minutes. He did not sit passively, but moved in a circle and worked his legs in and out

> of the jelly. The egg mass increased in size with this treatment, suggesting that the male provided moisture from his skin.
>
> Males can care for more than one clutch at a time. I saw the captive male sitting on both a fresh clutch and a hatching clutch in the same hour. [In another case a male left a spot where eggs had been laid. After two hours the male returned.] . . . He appeared to search for the eggs in the leaves for twenty-seven minutes, but was unable to find them, possibly because I had disturbed the leaves. He moved up the slope and disappeared into the litter five meters from the first site. I moved leaves aside and found him sitting on a clutch with three or four hatching tadpoles.
>
> . . . When eggs were nearly ready to hatch (after ten to thirteen days), the male sat on them, . . . with his back arched and his ankles pressed together, forming a sort of trough. The male never [sat this way] on freshly laid eggs.

When the tadpoles hatched they wriggled up the "trough" and tried to attach themselves to the male's back. Some have said that males of these species can carry more than one tadpole, but Wells never saw more than one tadpole on a male's back. The male then hops away, looking for a puddle, perhaps in a tree trunk, in which to leave the tadpole.

Male parenting is practically the rule in birds—found in probably 90 percent of the species making up the order. In a few dozen bird species the pendulum has swung over on the other side and the male provides most of the parental care. One theory has it that this happens when egg and/or clutch predation is especially intense. If the young are constantly being taken then it might be adaptive for both the male and the female for her to be ready to lay a new clutch at a moment's notice, which would involve her being up and around and feeding instead of sitting on the nest.

One of the best-known cases of role reversal is the American jacana (*Jacana spinosa*), a tropical shorebird that lives in freshwater swamps and marshes and feeds on insects. Females can have "harems" of males—stable pair bonds with more than one male simultaneously (the average "harem" size is 2.2). Males defend feeding territories within which they find their insects. They defend these territories against both male and female jacanas, but since the females usually weigh some 75 percent more than males (and are dominant over them), males usually succeed only in repelling invaders of their own gender. To drive out females they depend on the resident female, who defends a superterritory that embraces the ranges of two or more males. The females not only help exclude invaders, but even keep the peace within their "harem" by preventing incursions of one member onto the territory of a second. "Such polyandrous females may copulate with a male a few minutes after driving him from his neighbor's territory," Donald Jenni, the foremost American student of these birds, has noted. (A consolation copulation?) The male scrapes together the nest, broods the eggs, and attends and defends the young after the eggs hatch. The young can feed themselves, but are reluctant to do so except in the presence of the father. Predation is in fact high; probably less than 50 percent of the clutches are successful, and the males and females both spend a lot of time threatening, attacking, and distracting a whole zoo of avian, reptilian, and mammalian egg and chick eaters.

The females often mate with each member of their "harem" each day. Jenni says they display "an exceptional amount of overt sexual behavior." The precopulation ritual consists of an initial display of female aggression, then male appeasement of the female, followed by female solicitation of a male mount (which the male refuses more often than not). Mating is suspended while a male is incubating and until the chicks are six weeks old, whereupon it resumes—weeks before the male nests again. (The jacana breeds all year.) The female life-style seems clearly to be the more hazardous or exhausting, since females are displaced by other females twice as often as males are by other males.

Paternal care among the mammals is not a great deal more common than among the fish. The most famous single instance is probably the male beaver, who helps build a dam, a lodge, and accumulate a central food horde that will be drawn upon during the winter by his whole family. Some of the social canids, like the wolves, wild dogs, and jackals, have well-developed fatherhood roles. An average pack of wild dogs might consist of four or five males, all brothers, and a single unrelated female. The female bears a large litter, of about a dozen pups, which the adults feed either by the males regurgitating food to the nursing female or by all the adults regurgitating to the weaned pups. There is even one known case in which a litter of Cape hunting dogs was raised entirely by the males in the pack after their mother died. Males in a family of monkey species that includes the marmosets and the tamarins often carry their young through the trees, groom them, and have been seen giving their young food, though it is hard to know how significant this behavior is. Paternal care has been seen in several dozen rodent species, of which the beaver is one example. Two zoologists who raised and watched some pairs of deer mice (*Peromyscus leucopus*) report that at feeding time the males often:

> . . . remained in or near the nest while the female alone occupied herself with the food supply. At such times the male might remain watchfully alert or he might, if near the young, pull the nesting material over them, hover over them, or even wash them. Often he remained with them for as long as five minutes while the female fed. . . .[75]

Studies with paternal-care rodent species have shown repeatedly that pups reared in the company of both their mother and father survive better and grow faster than those reared by their mother alone. One such study, done with a predatory, highly aggressive desert mouse (*Peromyscus californicus*) found that daughters tended to be more active and exploratory, and sons more aggressive, if they had been raised by both parents. Progeny raised in single-parent families were also slightly less efficient at killing and eating crickets, an important food source for these rodents. The researchers suggest three reasons why a paternal presence might have these effects:

having both parents in the nest means an increase in the number of direct, physical contacts and stimulation; if the male and female parent swap periods of nest attendance, the nest temperature will be higher and more stable than if one parent had to come and go; and perhaps prolonged exposure to male pheromones might have some effect.

Whenever these different forms of male parenting slow down the male reproductive act, they bring the cycles of the two genders toward a common rhythm, into synchrony. The rate at which males rejoin the mating population after reproducing falls drastically. The population of competing males shrinks and females lose their leverage.

One group of animals in which this can be seen is the sticklebacks—small, temperate-zone and subpolar coastal and river fish. They are carnivorous and eat a wide variety of worms, daphnia, cyclops, fly larvae, and small crustaceans. One of their favorite foods is fish eggs, particularly the eggs of other sticklebacks. The males have taken on the task of protecting the eggs fertilized by their sperm from the appetites of their cannibalistic colleagues. They do this by first building a nest out of vegetation, gluing it together with a kidney secretion, and then shaping it properly. A female arrives (we will return to the courtship procedure later) and lays her eggs in the nest. The male fertilizes them and flattens the egg mass into a sheet against the bottom of his nest. He then reshapes, repairs, and lengthens his nest so that the next egg clutch will no more than partially overlap the first one, like shingles on a roof. Males can fertilize, flatten, and overlap up to seven clutches over a one- to two-day period.

After a few days the male begins to aereate and ventilate the eggs by fanning water through the nest. He inspects his clutches constantly and will pick out and eat any eggs that die and become moldy. As development progresses and the eggs generate more metabolic heat, the male makes holes in the roof of the nest, almost certainly to enhance the nest's ventilation. Eventually he tears the nest apart and leaves the material piled up in a tangle in which his newly hatched fry secrete themselves. (The eggs hatch after seven to eight days.) Fry that swim away from the nest get sucked up by the male and spat back into the nest pit. Males guard their fry in this way for about a week after hatching. It has been found that sticklebacks raised without a father are extremely timid compared to normally raised sticklebacks. "Overfathered" fry, fish kept in the presence of their father for longer than normal, are much bolder than normal fish, or at least less responsive, when a predator like trout is introduced into their tanks. Facts like these may be taken as a kind of indirect testimony to the effectiveness of male sticklebacks in providing protection for their young.

Stickleback courtship is unusual in the distinct, if small, role that females are called upon to play. Both sexes develop courting colors. The males' are more conspicuous: their throat and forebelly turn red; their irises, green; and their back develops a greenish tint. But females also begin to show speckles of dark pigment on their abdomen. These spots make her abdomen glitter slightly.

When a female is ready to spawn, she leaves her school, out in deeper water, and swims toward the males' territorial areas. When a male responds he does so by approaching the female in a characteristic, zigzag dance; she "replies" by floating vertically, head up, and orienting her abdomen toward the male. This makes her silvery, speckled belly, already swollen with eggs, especially prominent. The male then "leads" the female back to the nest and points to the entrance with his body. She pushes into the entrance and the male then induces her to spawn by butting her gently on the flanks. If the female is very ripe she can skip the courtship sequence altogether and simply swim straight into the nest. In fact stickleback courtship is extremely flexible in all respects. One of the leading students of this fish, R. J. Wooten, describes the courtship sequence, much as I have given it here, though in more detail, and then says, ". . . occasionally a courtship is seen which almost exactly follows this idealized sequence."

Which really makes one wonder just how much the "idealized sequence" itself is worth.

Now why should the female decorate her abdomen and display it to the male? And why should he be interested? As female sticklebacks grow, the size of their spawn increases by an order of magnitude, from twenty to thirty eggs to three hundred to four hundred eggs. Obviously a male would prefer to mate with a female carrying a larger set of eggs. But why should his preferences be given any weight? Why should females be selected to recognize and respond to an aspect of the male value system? The only reason plausible to me is that sometimes female sticklebacks have to compete among themselves for males, and the females that were capable of advertising their virtues properly attracted the right kind of males sooner than those that didn't, and that difference mattered.

Male parenting brings the male and female cycles into synchrony both by slowing down the turnover of males and speeding up that of females. A large female stickleback who is lucky in her hunting for food can spawn every four or five days over a sixty-day breeding season—if she is able to find males when she wants them. Further, from a female's point of view, not all males are equally good father material. The larger a territory a male can maintain around his nest, the less he will suffer from egg-robbing by other sticklebacks. Even if a male defending a small territory does not actually lose any eggs, the efficiency of his fanning declines since he is so distracted by the presence of neighboring males. A female might therefore be expected to prefer, if she has the choice, males formidable enough to keep their neighbors at bay.

However, stickleback males, attractive or no, can brood no more than seven clutches at once. As a rule one would not expect this to be an important limitation on female reproductivity; in nature, on the average, males seldom reach their limit. But it would be surprising if they never did. When food is ample a large female can spawn two to three times faster than a male can run through his nest cycle (males build up to five nests in a season, according to observations made under laboratory conditions, which often bring out optimum performances). It seems perfectly plausible that some males, some of the time, will find themselves in a position to pick and choose whose eggs they are going to brood. Once this happens female attractiveness suddenly becomes an issue, and females would evolve makeup with which to decorate their abdomens. To attract males the females obviously have to communicate in the terms males are interested in, and what stickleback males care about are lush, ripe bellies, stuffed with eggs, and promising a wealth of paternity.

Male leverage appears at other points in stickleback mating behavior. Occasionally populations of sticklebacks are found in which the males have abandoned their green and red courting colors and are a cryptic black or drab silver. In one lake in Washington, only 15 percent of the samples taken were "normally" colored. In laboratory experiments it has been clearly shown that if females are offered a choice of males to approach, they prefer the red and green ones. Obviously in these populations the predators had sufficiently decreased the population of adult males so that five out of six (in the Washington example) were able to ignore female preferences altogether! In most species, needless to say, males can't get away with doing things like that. If a certain color scheme is important, either in systems of female choice or male competition, then the males have to stick to that scheme, regardless of the consequences. An inconspicuous male might lead a long life, but it would be a celibate one. In the case of the sticklebacks, though, the males have more than their usual influence. If predators reduce their numbers even a little, which further increases the leverage of those that remain, the combination might be enough for males to get away with denying females whatever services bright courting colors perform for her. (Prediction: I bet those males that carry courting colors in generally cryptic male populations will be shown to be unattractive in some other way. They probably have small territories and so have little to offer a female but flashiness.)

One of the clearest indications of male leverage is females fighting with each other. In the *Neanthes caudata,* a bottom-dwelling marine worm in which the male incubates the eggs, females are intensely, even mortally, antagonistic of each other, but tolerant of males. Male emperor penguins incubate the pair's egg over the long Antarctic winter inland, while the female returns to the sea to feed. Whether a male will succeed in this strenuous chore (chill factors can fall down to a hundred below and more) depends on his fat stores; if a male gets too thin he will simply kick the egg away and try to get back to the ocean. A big, fat male is thus a creature of great worth and females have been seen to fight with each other for males with, as the French biologist observing them has said, "*la grandeur.*"

What appears to be male incitement of female competition has been witnessed in Wilson's phalarope (*Phalaropus tricolor*), a graceful, sandpiperish marsh bird that breeds in aggregations on prairie wetlands and Arctic tundra. The males provide almost all the incubatory care. Most hens breed once and have weak pair bonds with their male, but some manage to find or attract more than one male who will accept their eggs, so that some females are excluded from breeding altogether.

Phalarope females are larger than males, more brightly colored, and highly aggressive among each other in the presence of a male. Females court males by attaching themselves to one and then following him about, all the while threatening or attacking any other female that approaches, or is approached by, the couple. In one series of observations courting females were found to spend 80 percent of their time in hostile interactions with other females. (Do the males deliberately test their females in order to acquire the best mate?) Often females will fly after any unattached male that appears. Sometimes males that are being pursued in this way will swoop down over swimming females, drawing them into the chase.

These males then stop and hover and the pursuing females fight among themselves in midair as he watches. "Hovering males almost always give 'Courtship *Ernts*,'" one ornithologist reports in describing their vocalizations, "the same call which provokes interfemale aggression on the water." [76]

Female fighting has also been seen in *Dendrobates*, the frog in which the male carries the tadpoles about on his back. (Male *Dendrobates* have also been seen fighting with each other, though perhaps that was over a different issue than mate competition.) *Dendrobates* females have been observed eating other female's eggs, which if it is a technique for freeing up parental males, is another form of female competition. A female button quail, a species in which the male does the incubating, has been seen attacking a clutch of eggs being brooded by a male. Male *Neanthes caudata*, the paternal-care marine worm mentioned earlier, tolerate females until they begin brooding, at which point they attack approaching females. If they do this because female cannibalism is a threat to their brood, then that cannibalism might be aimed at freeing up the male as much as acquiring some rich protein stores. Both, in this system, are valuable resources, and infanticide might be selected for exactly the same reason it was among langurs.

Female fighting is not as common as a greater willingness by the females to pick up some of the burdens of the mating interaction. Robert Trivers, who more than any other biologist has alerted naturalists to the importance of "male parental investment," has pointed out that female courting, and bright female colors, are found in some species of seahorses and pipefish. Wild dog females seek out and court the male "brotherhoods" of wild dog packs. Kentwood Wells writes, describing the courtship of the *Dendrobates* frogs:

> Courtship began with a female approaching a calling male and sometimes touching his snout with hers. The male would then move off through the leaves with the female close behind. . . . As a pair moved through the leaves they engaged in bouts of tactile courtship. In every instance, the female played the

> more active role in these close-range interactions. She would jump on the male or place her front feet on his back and gently prod his back and vent region. Sometimes the female would climb across the male or sit on him and drum her hind feet on his back. . . . I never saw a male climb on a female's back. . . . Sometimes a female would crouch in front of a male and encircle him or rub her head on his chin. . . . As the pair moved through the leaves, the female usually became more active in jumping on the male as courtship progressed.

Sometimes a male can be seen leading a whole suite of females behind him.

Wells believes that what the female is doing is communicating her schedule to the male so that he will know when her eggs are ready to be deposited. She is, as it were, counting down for him, just the way it is done at Cape Kennedy. From the male's point of view, this allows an efficient mating; he will not be wasting his time with females that are still involved in the preliminaries of their schedule. Generally all males would like this information; there are a number of mammalian males that acquire it by tasting and decoding elements in the female's urine—but they have to do all the work themselves. Only the *Dendrobates* males can force the females to tell them explicitly. The reason (Wells thinks) is that their mating system evolved in a context of female competition: Males were in a position to pick those females who made matters easiest for them.

Males of the bird-of-paradise family (peacocks, bowerbirds) are usually promiscuous and offer no parental care. They are famous for their elaborate courting plumage. But there are three species (genus *Manucodia*) in this family in which the two sexes form pair bonds and the males help raise the young; in these species the sexes look alike, which suggests a more equitable distribution of these burdens. Pigeons are a monogamous species in which males parent. In fact, male pigeons even have the ability to make milk, in a gland in the esophagus, with which they feed their young. (Male ring doves and emperor penguins have the same capacity. So do female pigeons.) Mate-choice experiments, in which male and female preferences were deliberately pitted against each other, showed that females settled for a nonpreferred male sooner than males settled for a nonpreferred female.[25] Again, this is an illustration of male leverage, or at least, substantial direction away from the usual mating situation, in which males are glad to get any female they can. Very few males, among species generally, are ever in a position to define and respond to issues of female attractiveness, of differences among females. These mate-choice experiments show that in a few species males can define and adhere to standards of mate selection—and they can do this because their household tasks keep them out of competition with each other.

Just about as many female deer mice (*Peromyscus leocopus;* the paternal-care rodent mentioned earlier) are caught in traps baited with male pheromones as males in traps baited with female scent. This shows that each gender is equally important to the other; each works equally hard to solve the who-where-when problems of mating. Courtship is dramatically evenhanded:

> The two animals chased one another, traveling in head-to-tail formation with either animal in the lead or with both animals simultaneously leading and following as they ran in small circles. The chasing was interrupted by intervals of standing on their hind legs, front-to-front with noses touching, and of nuzzling one under the body of the other. As they moved under one another there was more naso-nasal touching and nose-to-genitalia exploration as well. They often turned over together, sometimes venter to venter, appearing to wrestle gently. . . .[75]

The two sexes "courted" not only when they were strangers introducing themselves to each other but after the birth of a litter, and after the female was returned to her mate from being removed for a laboratory procedure of some kind. Males did "call" more than females, which is believed to represent a slightly greater effort toward mate attraction.

There are a number of species that are monogamous, in

which the sexes should have equal degrees of influence over each other, but which look like female-leverage species. Pigeons are an example; the male takes the initiative in pair formation, does most of the courting and has slightly more colorful feathers, despite his paternal nature. There are three reasons why one might see this. First, there might be fewer females than males, so that males are in a poor power position regardless of how much they contribute to the offspring.* Second, the monogamy might be an illusion. Males might be able to have more mates—perhaps through having a longer reproductive life—than females, and so will come into a polygamous competition with each other, regardless of how paternally they act toward their progeny. Or the males might be mostly monogamous but competent, if conditions alter, to take up polygamous habits. An example in the cichlids, a fish known for its monogamy, pair-bonding, and male parenting (in some species both parents produce a nutritious slime on their body surfaces from which their fry can feed when hungry). Usually the male role is centered around protecting the young from predators, especially other cichlids. Sometimes a male that is already mated to one female may be seen courting a second female as he patrols the borders of his territory. Evidence like this suggests that if a male perceives a decline of interest in the local predators, a lessening of the threats to his own brood, he will then try to acquire a second mate. Cichlids look alike but the male is larger than the female. Larger males are not the usual pattern in fish; the pattern hints at a certain degree of male competition.

The third possibility why an equal-partner species might look like a female-leverage species is completely untestable, but ought to be at least mentioned. It raises the thorny question of what sorts of units evolution works with. Suppose that

* There are a number of species of hole-nesting water fowl in which the males are not colored cryptically while the females are, despite these being monogamous species. Entirely apart from the possibility that camouflage may be less useful to a male (who doesn't nest in the hole) several studies show that females in many of these species are differentially vulnerable to predators, which gives the survivors leverage.

the behavior or structures that we now see being used in courtship originally evolved because of their usefulness in some other context—like territorial defense. Then it might be that these displays could be drawn upon for courtship purposes just because they were available, in preference to developing another vocabulary. In that case while it might appear that males of such species were taking the lead in courtship, the "reality" is that the world has trained them to act out their feelings in more expansive and energetic terms than the females. But if we could assess the emotional states of both animals directly we would find that they had the same attitude toward each other, whatever that attitude might be.

One route by which male parenting is thought to have evolved is via that ancient evolutionary dynamic—female choice. Females, this theory proposes, have some way of knowing which male will make the best parent and they have exercised their leverage to select for that quality. The paternal behavior for which this idea works best is female feeding, bringing the female a meal which is then turned into egg protein. This can be an important source of food; one ornithologist found that a female blue tit that begs food from her mate (when her mate approaches with a piece of food she stops foraging, shivers her wings, and calls) gets 250 percent more food than does a solitary, self-sufficient female that never begged from males. (The reason for the large return to begging is that the food items brought by the blue tit males are larger than those taken by females foraging on their own.)

It seems easy to imagine how this kind of parenting might have grown out of female choice. There are many species in which males use food to attract the female during courtship. William Calder of Duke University described such a courtship among captive roadrunners (*Geococcyx californianus*):

> A mouse given to one of the males often seemed to act instantly as an aphrodisiac. . . . Bearing this food, the male went to the female, usually approaching from the rear. Sometimes the female begged like a roadrunner chick, fluttering her wings and uttering a buzzy, squeaking call.

> The male then [raised] his crest feathers . . . [until] the colored areas of skin posterior to his eyes were maximally exposed. He wagged his cocked tail from side to side, while rapidly stepping or patting his feet in place. A rapid, vocal *kuk-kuk-kuk-kuk* accompanied this. After a short period of wagging and stepping he made a deep bow in which the tip of his tail nearly touched the floor. . . . A low, almost growling *coo* was made during the bow.

All this time the male continued to hold the mouse in his beak; not until the copulation was over was the food given the female. While, in this specific case, the mouse probably served as a cosmetic, as a way of riveting the female's attention to the male, it is easy to imagine this behavior ending up as mate-feeding, blue-tit style, assuming only a context of male competition.

One species that has gone a little further down this road is the eastern scorpion fly (*Bittacus apicalis*), which has been the subject of a masterful study by Randy Thornhill of the University of New Mexico. The scorpion fly lives in the lush, dense undergrowth of the moist, deciduous forests in eastern North America. There it hatches, hunts (by pouncing on other flies, like house flies, in the air and then sipping their fluids), breeds, and dies in the early summer. Both males and females hunt for the first two or three days after they emerge, but when breeding season begins the females stop hunting and are fed entirely by the males. When a male captures a fly he first tastes it briefly, and then either discards it and hunts down another, or lands on a twig, elevates his abdomen, and calls for a female by emitting pheromones. If a female responds she lands beside the male and lowers her wings. He presents both the fly and his genitals simultaneously. The female tastes the prey—Thornhill calls it the "nuptial prey item"—which continues to be held by the male, and either feeds from it or flies away. Only the female feeds during the copulation.

Thornhill found that the flies discarded by the males tended to lie in the smallest third of the prey-size range. He also found that the length of the copulation, which could proceed for as long as twenty-five minutes, seemed to correlate directly with the volume of the "nuptial meal." When a male tried to present a female with a fly that measured less than 20 mm^2, she would either refuse to copulate at all or break away after just a few minutes.

> One male presented a [fly] 10 mm^2 to an attracted female. The female began to feed on it, but each time the male attempted to copulate with her she pulled her abdomen back, thereby preventing copulation. After about one minute the male pulled the prey from the grasp of the female and presented it to her again. The female again refused to copulate and after another minute the male grabbed the prey from the female and the pair parted. . . . within four minutes [the female] had coupled with another male which possessed a nuptial meal of 20 mm^2. This copulation lasted twenty minutes and was terminated by the male. In the second reproductive encounter the female showed no signs of coyness and coupled with the male after tasting the nuptial meal.[133]

Thornhill found that for the first five minutes or so of the copulation the females prevent any sperm from entering their body. After five minutes the number of sperm permitted to enter climbs rapidly for twenty minutes, after which the increase flattens out. At this point the male usually breaks off relations. The male and female fight over what remains of the fly, if anything does. If the male wins he examines it once more, and then either calls for a second female or discards it and begins hunting again. The female then retires and immediately lays an average of three and a half eggs.

There could hardly be a clearer illustration of the influence of female choice on the evolution of male parenting. Yet logic argues that as important as female leverage is in other contexts, it can have only limited usefulness in forcing the evolution of male parenting. The trouble with male parenting is that it takes time, and reproductive acts that take time destroy the whole basis on which male competition is built. The scorpion-fly female lays her eggs over a four-hour period, during which time she is unresponsive to other pheromone calls, and after which she becomes receptive again and rejoins the mating pool. She might mate four times in a day. Males need about fifty minutes to find a prey of an acceptable size (to establish this point Thornhill individually marked forty-two males and followed them, stopwatch in hand, as they hunted), and this limits their mating opportunities to about ten a day. Thus, assuming that the females are so efficient in finding males that they enjoy their maximum possible advantage, their edge will be four to ten or one female to two and a half males. This still puts them in a powerful position, but there are many females in nature that play from a much stronger hand.

The scorpion-fly sex system is rich with signs of emerging male leverage. First, the male does not go looking for the female; she flies to him. He does not even have to distinguish the genders; the female identifies herself. In fact, a whole class of scorpion-fly male parasites have evolved that make use of this male dependence on female courtship. These "transvestite" males, as they have been called, follow up pheromone calls, land next to the calling male, lower their wings just as the female does, and grab onto the fly when the male extends the prey. Of course they do not couple; they keep their genitalia just out of the reach of the calling male. After a few seconds it usually dawns on the calling male that something is wrong and he then tries to snatch the fly back. But by this time the transvestite has gotten a firm hold on the fly and sometimes suceeds in wrenching it away from the male that originally caught it. The robber then flies off and uses the stolen fly as bait for females for himself.

Male parenting need not, however, depend on female choice and male competition over numbers of mates to drive its evolution. A hunting insect like a scorpion fly cannot conceal itself from predators as easily as can a vegetarian insect, which can hide under leaves or burrow into stems. Thornhill does not describe the intensity of the risk that scorpion flies undergo from birds while they cruise about looking for game, but since they live in a bird-rich habitat, and have their population peak during a period when many birds are breeding, there is nothing implausible in assuming that the risk is high. If so, then males that fed the females that they mated with, and thereby allowed them to keep out of sight while laying those eggs that carried the males' sperm, might outcompete males who do not give gifts and force their females to fly around hunting for protein. The risk to females, and therefore the gain to males who kept their mates out of sight, would have to be large, but perhaps it was large. In other words, depending on the situation, male competition alone can drive males into forming cooperative relationships with specific females. In the situation envisaged here the scorpion-fly males are not (only) competing over the number of females each can fertilize, but over who can get the most potential offspring past that critical moment when the mated female is vulnerable to predators. Depending on how important *that* competition is, females will not need to exercise any choice at all to get themselves a provider. In fact, to push the argument to an absurd extreme, males in such a system might be selected to force-feed females if necessary, not that it ever would be. In this context male parenting can evolve even when the pool of competing males is comparatively small; female leverage is just not relevant.

Much of this book has tried to imagine ways in which the apparent paradox of the nonproductive male might be seen as making evolutionary sense. To the degree that such ideas seem plausible they raise questions about *productive* males. What causes males to give up chasing and competing for females? And why should females ever be selected to give up their leverage and put themselves in a position to have to respect male needs and values on the who-where-when-how issues?

It is easier to see what causes males to "settle down." Paternal males, where they exist, usually play critical roles in the reproductive cycle, and it is easy to imagine that a promiscuous male, under those conditions, would lose all his offspring, no matter how many he fathered. The burying beetles sequester large pieces of food that are usually fought over by several beetles. Any male that fertilized a female and left without helping defend and bury the carrion would have no prospects at all. The cichlids and the sticklebacks live in environments that are fiercely hostile toward their progeny; parent-removal experiments show that the fry get eaten up within minutes after they have lost their parental protection. (I know of no experiments in which only the male was removed.) Male parenting is also found in many tree-dwelling species—many birds, and some arboreal primates, in which the males carry the young. It seems plausible that in these species it is important for the female to be light, and that value selects in turn for early deliveries, for her young being borne long before they are large enough to take care of themselves. The addition of male care might make an especially important difference in these conditions. Among birds generally a lack of male parenting seems to be associated with food that does not need to be caught. Promiscuous species tend to eat nectar and fruit, which, when available at all, is abundant; polygamous species tend toward seeds and plant food; while monogamous birds tend to eat insects and other animals. It may be that in flesh-feeders the amount of food returned to the nest can make a big difference in the number of offspring successfully fledged, while in the other species the young already get enough food; what limits their number is some other factor (such as the brevity of the fruiting season) about which males can do nothing. In some species of flesh-eaters, like gulls, one parent is needed to stand guard at the nest against the attacks of other members of the species.

Even if a male doesn't play a critical role in reproduction, he will be selected to contribute to his offspring if he has nothing better to do anyway; if circumstances prevent his competing over more females. This might happen if the females are dispersed widely, or pass into and out of estrus too quickly, for him to be able to visit more than one. Or it might be too dangerous for him to move around. The elephant shrew (*Elephantulus rufescens*) is a small insectivore that has adopted, as its main antipredator strategy, flight. It maintains a maze of cleared pathways within its territories down which it zigs and zags when being chased, for example, by a snake. Males and females share a joint territory and nothing else. They do not sleep or rest together, or court, feed, or even groom each other. Since the males do not have to maintain the level of protein intake the females do, they have, relatively, some time free, and they use that time to keep the system of runways clean. They spend twice as much time maintaining the escape routes as the females. While it might be argued that this kind of male parenting could only make a slight difference to the male's reproductive success, it would make more of a difference than anything else he might do with his time—he's certainly not going to abandon his runway network to roam around looking for females—and that's what counts.

So one can imagine selected, special cases in which male parenting might be adaptive for males, but what about females? Why should they prefer goods to service? Why suffer the inconveniences that attend their giving up their leverage over males? In general the freedom of choice seems to be important to females; why should they be willing to give it up? Some females do in fact reject male parenting; there are species of marsh birds in which the males do feed the young and perform other functions. Yet even so the females sometimes prefer to nest polygamously, with two or so females to a territory. This dilutes the amount of care the territorial male can give, but if his territory is rich enough the females will prefer to live there rather than accept the assistance of a male in a poor territory. Studies have shown that the territories of bigamous males are both larger and contain more standing water (wrens feed on aquatic invertebrates). They also have more male-built nests; the investigators suggest that the number of nests a male builds is an index of that male's territory's quality, since a male that could build many nests obviously needed to spend very little time looking for food.[145]

This is a good example of the advantage to females of a polygamous system. The good territories are open to more than one female, and the males have been selected, by virtue of their competition, to advertise the quality of their territories in a convenient, easy-to-read, summary form. Thus the females can choose, resources or paternal care, whichever promises the best results.

Male parenting is thought to evolve in environments that are ecologically stable and (therefore) socially competitive, habitats in which excess population growth is cut back by competitive interactions among the members of the species. These are the circumstances in which each new increase in male care is likely to be strongly rewarded, since very small initial advantages in size and/or training of the young can determine who wins these encounters, and do so over and over. A small edge early in life can, if the rules stay the same, snowball to a large payoff by adulthood. Male parenting allows the young to stay young longer, and therefore to bring more to the business of being adult—whether a larger size, a more developed nervous system, or a better sense of the environment—when they do finally go off on their own. The examples of male parenting given here have tended to stress the protection of vulnerable young, and certainly when one sees progeny being threatened by predators several times their size, that seems the most sensible interpretation. But at least in some cases, especially in some bird and mammal species, one wants to ask why the young are born so vulnerable in the first place and then take as long as they do to reach adulthood. It can be argued that when all the juveniles in a given generation are going to be forced to compete with each other over a very limited number of nest sites, or foraging territories, when the main barrier between them and successful adulthood is not huge predators but each other, then prolonged immaturity and, consequently, vulnerability might actually be selected. Offspring tolerated by their families for any length of time are so treated because it is to everyone's advantage that they learn as much as possible about the world and how to live in it. One way of giving them a "head start" is to get them born and into an environment that is both more open to the world, and still protective. If there are any irrevocable decisions that have to be made about how adult life is to be entered and practised, then it is better to postpone these until adulthood is in fact at hand and the conditions of the moment can be examined. Schelling, the nineteenth-century German romantic philosopher (who wrote on biology), believed that the position an organism occupied on the scale of life, the great chain of being, depended on how long it had managed to delay sexual maturation. Once a creature became sexual its fate was sealed, Schelling thought, or at least defined. Lower organisms were those that had succumbed to temptation and indulged in sex at an early age. While they may have enjoyed themselves at the time, they paid for it, because they were forever frozen out of the chance to ascend to the higher levels of organization, and had to trudge through life as a worm or a plant. Organisms with the self-restraint to postpone sex, postponed with it its rigidification and commitment.

While Schelling's ideas in their original form have little but entertainment value for the contemporary reader, they do hold a grain of truth, which is that keeping one's options open till the last possible minute makes good sense. When paternal care makes this possible, then the association between male parenting and offspring dependency will be strengthened. Finally, paternal care allows progeny to develop adaptations that need practise and polish to be useful. One might imagine, looking at the halting, callow, incompetent manner in which a young fox pup hunts, that male parenting evolved because fox puppies are so vulnerable and dependent—but in fact it is just as easy to argue that a protracted period of dependency is a great plus in evolution, that foxes could never hunt as well as they do without having a part of their lives devoted explicitly to studying and practising the craft.

Earlier it was argued that both size and specialization,

including the specialization of gender, are adaptations to social competition, to living in a species that prunes back its own excess population growth. The advantages of size flow from its usefulness in direct competition, "when push comes to shove"; of specialization, in the indirect competition of efficiencies. A specialized forager will crop its habitat faster and closer than a less specialized peer. It is possible that male parenting is a way of enhancing both these modes of social adaptations at once, or alternately. A period of protected immaturity allows the young freedom from worrying about the two priorities of adults—survival and reproduction—and they can, instead, put all their energies into learning and growing.

From the female's point of view male parenting might well be a net benefit, but it is hard to believe that there are not some costs to her in accepting it. One would think that the mating system would have to become more sluggish and less convenient. She has become dependent on another creature for an important part of her reproductive cycle. The male might compete with her for resources. In many species the female mates repeatedly with the same male, and this will clearly restrict the genetic variability of her offspring. And male-parenting systems generally require females to be more sensitive to male needs, because the pool of competing males is less likely to be a factor.

The silver lining in all this, one supposes, has to be male care. There might be a second advantage, springing, ironically, out of one of the apparent disadvantages of male parenting, which is the dependency of the female on the male. Seriously dependent relationships, apart from those of the young on the parent, are uncommon in nature, at least among sexual animals. Symbiotic, mutual-benefit relationships are not rare, of course—the birds in a flock increase their foraging powers by combining, or starlings amplify their defensive powers by flying in groups (they bunch together when a hawk flies overhead; this makes it harder for the hawk to strike at one starling without inadvertently colliding with a second, which, at the velocities hawks reach in their dives, would be likely to seriously injure the raptor). But these intraspecies symbioses tend to be fairly loose; they are neither highly elaborated nor do they turn on the competence of specific individuals. The ordinary sexual relationship is an example. Females are not dependent on males for anything but the performance of a routine that can be acted out by all males interchangeably (and they might accept multiple fertilizations in addition); successful males mate with many females, which makes them less dependent on the luck or competence of any single one of their mates.

Male parenting involves a quantum leap in the degree of interdependence experienced by the partners. The females in such species depend on specific males to perform their paternal role, while males commit their reproductive prospects to a small number of females, perhaps to even just one. There are exceptions. The stickleback and the scorpion fly combine male parenting and high degrees of mutual independence by breaking up the breeding season into many reproductive cycles. A third example is the American rhea, a ground-nesting Latin American bird. In the spring the males compete among themselves to acquire "harems" that range from between two to fifteen females. Each male builds a nest, and all "his" females lay in that nest until the clutch has built up to a few dozen. Then the male (apparently) drives the females away and begins to incubate the eggs. The females go off to consort, copulate, and lay eggs in the nest of a second male, and might in a single season lay in the nest of as many as seven males. While the rhea mating system may not have evolved for this reason, it does allow each bird to hedge its bets and spread the risk of failure.

As a rule, though, male parenting is more entangling, and a number of experiments have shown that this increase in vulnerability is serious enough to have evolved its own protective devices. David Barash, a psychologist-zoologist at the University of Washington, reports that when female mallards are "raped" by strange males (the mallards are one of the very few species in which rape has been observed), the female's mate intervenes aggressively about one-third of the time, usually by grabbing the assailant by the neck and beating him with his—the male mate's—wings. (One reason why the males intervened only one-third of the time was that mallard rapes are usually committed by multimale gangs; the mated males were only half as likely to intervene when their female was being assaulted by a gang as when she was attacked by a single male.) After the rape about one-third of the mated males copulated immediately with their mates, skipping almost entirely the usual precopulatory displays. Barash believes this is an anti-cuckolding trick; an attempt by the mated male to compete with the sperm introduced by the rapist. Mallards are essentially monogamous; males are dependent on a single female to bear their progeny. This dependency leaves them vulnerable to cuckolding and makes it necessary for them to defend their mates. Barash believes that one reason why these hasty copulations did not follow rapes in every case is that mallard females can tell whether sperm was transferred during a copulation and so can inform her mate whether or not a rape did constitute a genuine threat to his paternity.

In another series of experiments, Barash placed model male mountain bluebirds (*Sialia currucoides*) on the nests of mated pairs while the male was out foraging. He found that when he did this before the female had laid her eggs the male attacked both the model and his mate (he "pulled an undetermined number of primary feathers out of his mate's wing . . .") and then drove her away, replacing her with another female. But when he placed the model on a nest in which the eggs had already been laid the male was much less hostile to the female.

Barash interprets all these results as adaptations by males to their being dependent for a major fraction of their procreativity on one female. The male mountain bluebird drove away the female that, appearances suggested, had been fertilized by another male, because if "his" female had laid eggs not fertilized by him he would have been completely shut out of reproducing that season. Male mountain bluebirds have leverage; they seize and defend nest sites, which are in short supply. The male bluebird could therefore afford to drive his adulterous female away. Male mallards have no leverage and therefore have to settle for the more problematical strategy of physical intervention and sperm competition.

A third experiment pointing to the same phenomena was done at Duke University with pairs of ring doves (*Streptopelia risoria*).[59] Two groups of females were used; one composed of female doves that had previously been courted by males and the other of females that had been isolated from males. Male doves were placed first with one group of females and then, a few days later, with the second. The females that had been courted already gave a more advanced response to the displays of the males than did the "unexposed" females. This advanced response had quite an effect on the mood of the courting males. Overall they were about twice as aggressive toward the "exposed" females and courted them only half as much. "The difference in response may be related to the differing probability of cuckoldry," the experimenters conclude.

These experiments stress the negative side of interdependence, the prospect, or reality, of the contract being broken. There is a positive side, too, and that is the incentive to capitalize on the prolonged associations often found between the sexes in male-parenting species to fine-tune the relationship. Ordinarily, when a successful male mates with many females and a female with many males, then any one mate is, for the other, always just one of a number. Time spent fine-tuning one of these relationships is time wasted. But when two animals mate together repeatedly, so that their procreativity is represented and registered by the same set of offspring, they will, from the point of view of evolution, be selected as though they were one bisexual organism. Each partner will be adapted to enhance the performance of the other almost as strongly as it will be to improve its own. (I say "almost" only because the loss of one partner and then pairing with a new one is likely to be part of any animal's experience, no matter how tenacious their pair bond.) If these partners associate with each other continuously, then they will have both the incentive and the opportunity to devise individualized, particularized accommodations to each other's qualities that act to enhance the reproductive efficiency of the pair as a whole.

The classic study of this effect was made by J. C. Coulson, a zoologist from the University of Durham in Great Britain, who studied a colony of kittiwake gulls (*Rissa tridactyla*) for eleven years. He found that females that changed their mates did not lay as early in the season, laid smaller clutches, and despite having fewer eggs to look after, hatched fewer eggs. "This suggests that the birds in a newly formed pair are less well adjusted to each other," Coulson writes, "that they do not stimulate each other to as high a level of reproductive drive, and that they cannot coordinate their incubation pattern as well as a long-established pair." To give an example, he found that a female that changed her mate between breeding seasons delayed laying her first egg by about four and a half days. Even having changed mates a year ago, before the last breeding season, she still delayed egg laying by about three days. A female that had changed mates for two successive years retarded the date when she laid her first egg by over seven days. These and other negative effects on the kittiwake breeding biology affected females regardless of whether their partner had died or been "divorced," and so were a product of beginning with a new mate rather than the reason for the change. Changing partners is not necessarily a bad idea; a pair that had hatched no eggs at all in one season might profit from a change in partners, and, indeed, two-thirds of the birds that failed to hatch a single egg did "divorce" and re-pair. Results confirming and paralleling

Coulson's have been found in the red-billed gull (*Larus novaehollandiae scopulinus* Forster) and the Arctic skua (*Stercorarius parasiticus*).

Work like Coulson's is the strongest argument that pairs can develop partner specializations that allow them to attack the various issues in their lives, including but not restricted to reproductive issues, in a coordinated way. There is much indirect evidence that also points in this direction; evidence that suggests that it is important for each member of a pair to communicate more or less continually with the other.

The best singers in birds, those with the most complex songs, tend to be monogamous. From one point of view this is a little surprising, since one economical way of accounting for complicated birdsongs is to suggest that they evolved under sexual selection, just the way peacock displays evolved. Yet the males of promiscuous and polygamous species, who are most likely to be subject to sexual selection, have comparatively simpler songs. Monogamous birds also have the best-coordinated and most precise songs. These are called duets, or duetting, in which mated pairs keep in contact by calling antiphonally back and forth, the first vocalizing one or more notes and its mate instantly responding with a variation of the first call. "So fast is the exchange," E. O. Wilson writes, "sometimes taking no more than a fraction of a second, that unless an observer stands between the birds or uses sophisticated recording equipment he does not realize that more than one bird is singing." Duetting is known in a great many birds, including cranes, sea eagles, geese, quail, woodpeckers, and in at least two species of monogamous primates: the siamang and the gibbon. The talking birds, like the mynahs, parrots, starlings, ravens, and others, are all monogamous. It is often observed of pairs in these species that each pair develops its own common call, distinct from those of neighboring pairs, through which they keep in touch. Several suggestions have been made as to the point of these communications. Many of these birds live in dense forests, and perhaps knowing each other's call allows the two birds to stay close to each other while foraging so as to share whatever discoveries either member makes. A second suggestion relies on many of these species being "opportunistic breeders," birds that can control their reproductive cycle so as to breed at moments that are optimal ecologically and physiologically. Such pairs might need both to stay in touch with each other and monitor each other's mood. A third explanation is that continuous and individualized contact reduces the risk of hybridization with similar-looking species. There surely can be no doubt that one of the virtues of the pair bond is that it does prevent hybridization; no hybrid plumages have ever been recovered from the three monogamous bird-of-paradise species mentioned earlier. But that is not to say that the pair bond evolved in the way it did because, and only because, birds with weaker bonds hybridized more often. A fourth possibility is that continuous communication allows each bird to compare the other's assessment of the environment with its own, so that an experienced pair might be able to forage, or detect predators, more efficiently in that each member would have access to the "judgment" of the other. The logic of this suggestion works best—the advantage of being an experienced pair is at its peak—if one assumes further that each bird can learn what the strengths and weaknesses of its partner are, and which responses are more likely to be useful, or right, compared with its own. If that can happen then each bird can learn from the other, can, as it were, grow in the relationship.

Whatever biologists finally agree on as the point of the pair bond, it adds an interesting final twist to the story of natural sexuality. Imagine distributing all living species into four groups, depending on how important social interactions are among them. The ideas discussed here would lead one to predict that the first group would be composed of creatures like bacteria, who are adapted to completely unpredictable and unstable environments, and who are seldom sexual at all.

The second group are those for whom social competition has become important enough for specializations to emerge, and therefore for parents to profit from an adaptation like sex that allows them to hand down a wider range of professional possibilities to their offspring than they could if they were asexual. But the competitive interactions are not as yet so important that there is any advantage to be gained in regulating and defining in depth how the issues raised by sex (who, when, where, how) are to be solved. These species are the bisexuals, creatures that are sexually generalized, that are all identical to each other.

The third point on the scale is reached when specialization reaches the administration of mating issues. Gender arises, and with it ranges of specialization and techniques of labor division that leave the sexes defined in very different ways.

The fourth stage is composed of those species that are so "well adapted" to their environment that they are seldom badly surprised by it. Whether they will succeed in breeding for a given season, or even over their life, is a social question, decided by social forces. In such species, large, well-informed young, whose skills have been brought to a high polish, will, almost by definition, have an advantage. This requires giving the young a protected environment in which to grow, and that need is both the opportunity and the reward for the evolution of male parenting. As male parenting develops, one force that was defining the sexes in opposite ways—their different reproductive tempi—ceases to apply. Moreover, to the degree that the protective environment must be continuously maintained, so that the two sexes have to spell each other, then a second force that was defining the two sexes in different directions, which was a role-segregated division of labor working within the mating interaction, also ceases to apply. As the two sexes become involved in the same tasks, as they do the same things, they will necessarily be selected in the same directions.

The result might look as though the species were becoming hermaphroditic again, but that is only because the unit of specialization has moved "up" one notch from single individuals to the pair itself. It is the pair that is, as a whole, selected to pursue strategies, maximize efficiencies, and interact as a unit with other pairs, other parents. The question is sometimes asked as to whether mated animals feel love for each other. Obviously this question can never be answered with real authority; but if, as many people think, love is a sense of dependency freely entered into and mutually maintained, then arguing that such animals as two old kittiwake gull mates do *not* feel love for each has to be judged an unwarranted and uneconomical speculation, with neither evidence nor logic to support it.

NOTES

1. Abele, Lawrence G., and Sandra Gilchrist. 1977. "Homosexual Rape and Sexual Selection in Acanthocephalan Worms." *Science*, 197:81–83.
2. Adler, N. T. 1974. "The Behavioral Control of Reproductive Physiology." In *Reproductive Behavior*, ed. by William Monagna and William Sadler.
3. Alcock, John, C. Eugene Jones, and Stephen L. Buchman. 1976. "Location before Emergence of the Female Bee, *Centris pallida*, by Its Male." *J. Zool. Lond.*, 179:189–199.
4. ———. 1977. "Male Mating Strategies in the Bee *Centris pallida* Fox." *Amer. Nat.*, 111:145–155.
5. Ames, Oakes. 1937. *Pollination of Orchids through Pseudocopulation.* Botanical Museum, Cambridge, Mass.
6. Armitage, K. B. 1962. "Social Behavior of a Colony of the Yellow-bellied Marmot (*Marmota flaviventris*)." *Anim. Behav.*, 10:319–331.
7. ———. 1965. "Vernal Behavior of the Yellow-bellied Marmot." *Anim. Behav.*, 13:50–68.
8. ———. 1973. "Population Changes and Social Behavior Following Colonization by the Yellow-bellied Marmot." *J. Mammal.*, 54:842–854.
9. ———. 1974. "Male Behavior and Territoriality in the Yellow-bellied Marmot." *J. Zool.*, 172:233–265.
10. Austin, C. R., and R. V. Short. *Reproduction in Mammals.* Cambridge University Press, 1972.
11. Ballard, Warren B., and Robert J. Robel. 1974. "Reproductive Importance of Dominant Male Greater Prairie Chickens." *The Auk*, 91:75–85.
12. Barash, David P. 1976. "Male Response to Apparent Female Adultery in the Mountain Bluebird." *Am. Nat.*, 110:1097–1099.

13. ———. 1977. "Sociobiology of Rape in Mallards." *Science*, 197:788–789.
14. Barlow, George W. 1974. "Contrasts in Social Behavior between Central American Cichlid Fishes and Coral-Reef Surgeon Fishes." *Amer. Zool.*, 14:9–34.
15. Bartholomew, G. A. 1952. "Reproductive and Social Behavior of the Northern Elephant Seal." *University of California Publications in Zoology,* 47(15):369–472.
16. ———. 1970. "A Model for the Evolution of Pinniped Polygyny." *Evolution,* 24(3):546–559.
17. Beach, Frank A., and Burney J. Le Boeuf. 1967. "Coital Behavior in Dogs. I. Preferential Mating in the Bitch." *Anim. Behav.*, 15:546–556.
18. Belding, David L. 1934. "The Spawning Habits of the Atlantic Salmon." *Trans. Amer. Fish. Soc.*, 64:211–18.
19. Bertram, Brian C. R. 1975. "The Social System of Lions." *Scientific American*, 232 (5): pp. 54–65.
20. Birky, C. William, Jr., and John J. Gilbert. 1971. "Parthenogenesis in Rotifers: the Control of Sexual and Asexual Reproduction." *Amer. Zool.*, 11:245–266.
21. Bonner, John Tyler. *Cells and Societies.* Princeton University Press, 1955.
22. ———. 1958. "The Relation of Spore Formation to Recombination." *Amer. Nat.*, 92:193–200.
23. ———. *On Development.* Harvard University Press, 1974.
24. Borowsky, Richard, and Klaus D. Kallman. 1976. "Patterns of Mating in Natural Populations of *Xiphophorus*." *Evolution*, 30:693–706.
25. Brown, Jerram L. *The Evolution of Behavior.* W. W. Norton, 1975.
26. Buck, John, and Elisabeth Buck. 1976. "Synchronous Fireflies." *Scientific American,* (May) 74.
27. Burley, Nancy. 1977. "Parental Investment, Mate Choice, and Mate Quality." *Proc. Natl. Acad. Sci. USA.*, 74:3476–3479.
28. Burns, John M. 1966. "Preferential Mating versus Mimicry: Disruptive Selection and Sex-limited Dimorphism in *Papilio glaucus*." *Science*, 153:551–552.
29. Butterfield, P. A. The Pair Bond in the Zebra Finch (pp. 249–276).

30. Calder, William A. 1967. "Breeding Behavior of the Roadrunner, *Geococcyx californianus.*" *Auk*, 84:597–598.
30a. Campbell, Bernard. Editor, *Sexual Selection and the Descent of Man. 1871–1971*. Aldine Publishing Co., Chicago, 1972.
30b. Charnov, Eric, and James Bull. 1977. "When Is Sex Environmentally Determined?" *Nature*, 266:828–830.
30c. Cohen, Jack. 1975. "Gamete Redundancy—Wastage or Selection?" In: *Gamete Competition in Plants and Animals*, ed. by D. L. Mulcahy. North-Holland Publishing.
31. Cole, Charles J. 1978. "The Value of Virgin Birth." *Natural History*, 87(1) (January): pp. 56–62.
32. Collias, Nicholas E., and Elsie C. Collias. 1970. "The Behavior of the West African Village Weaverbird." *Ibis*, 112: 457–480.
33. Coulson, J. C. 1966. "The Influence of the Pair-bond and Age on the Breeding Biology of the Kittiwake Gull *Rissa tridactyle.*" *Journal of Animal Ecology*, 35(2):269–279.
34. Cherfas, Jeremy. 1977. "The Games Animals Play." *New Scientist*, 75 (1069) (15 Sept.): pp. 672–673.
35. Croll, N. A. *The Ecology of Parasites*. Harvard University Press, 1966.
36. Curio, Eberhard. *The Ethology of Predation*. Springer-Verlag, 1976.
37. Cox, Cathleen R., and Burney J. Le Boeuf. 1977. "Female Incitation of Male Competition: A Mechanism in Natural Selection." *Amer. Nat.*, 111:317–335.
38. Crews, David. 1975. "Psychobiology of Reptilian Reproduction." *Science*, 189:1059–1065.
39. ———. 1977. "The Annotated Anole: Studies on the Control of Lizard Reproduction." *American Scientist*, 65:428–434.
40. Crews, David, and Ernest E. Williams. 1977. "Hormones, Reproductive Behavior, and Speciation." *Amer. Zool.*, 17: 271–286.
41. Cuellar, O. 1977. "Animal Parthenogenesis." *Science*, 197: 837–843.
42. Darwin, Charles. *On the Origin of Species*. (A facsimile of the first edition.) Harvard University Press, 1964.

43. Dawkins, Richard. *The Selfish Gene.* Oxford University Press, 1976.
44. Davey, K. G. *Reproduction in the Insects.* W. H. Freeman and Co.
45. Davies, N. B., and T. R. Halliday. 1977. "Optimal Mate Selection in the Toad *Bufo bufo.*" *Nature,* 269:56–58.
46. Davis, J. W. F. 1976. "Breeding Success and Experience in the Arctic Skua, *Stercorarius Parasiticus.*" (L.)*J. Anim. Ecol.,* 45:531–535.
47. Davis, J. W. F., and P. O'Donald. 1976. "Estimation of Assortative Mating Preferences in the Arctic Skua." *Heredity,* 36(2):235–244.
48. ———. 1976. "Territory Size, Breeding Time and Mating Preference in the Arctic Skua." *Nature,* 260:774–775.
49. Dewsbury, D. A. 1967. "A Quantitative Description of the Behavior of Rats during Copulation." *Behavior,* 29:154–178.
50. Dewsbury, D. A., and D. Q. Estep. 1976. "Pregnancy in Cactus Mice: Effects of Prolonged Copulation." *Science,* 187:552–553.
51. Downhower, J. F., and K. B. Armitage. 1971. "The Yellow-bellied Marmot and the Evolution of Polygamy." *Amer. Nat.,* 105:335–370.
52. Downhower, Jerry F. 1976. "Darwin's Finches and the Evolution of Sexual Dimorphism in Body Size." *Nature,* 263:558–563.
53. Dunn, Emmett R. 1941. "Notes on *Dendrobates auratus.*" *Copeia* (July 8): pp. 88–93.
54. Eaton, Theodore H., Jr. 1941. "Notes on the Life History of *Dendro bates auratus.*" *Copeia,* (2) (July 8):93–95.
55. Edmunds, George F., Jr., and Donald N. Alstad. 1978. "Coevolution in Insect Herbivores and Conifers." *Science,* 199:941–945.
56. Elliott, Phillip F. 1975. "Longevity and the Evolution of Polygamy." *Amer. Nat.,* 109:281–287.
57. Emlen, Stephen T. 1976. "Lek Organization and Mating Strategies in the Bullfrog." *Behav. Ecol. and Sociobiol.,* 1:283–313.
58. Emlen, Stephen T., and Lewis W. Oring. 1977. "Ecology,

Sexual Selection, and the Evolution of Mating Systems." *Science*, 197:215–223.

59. Erickson, C. J., and P. G. Zenone. 1976. "Courtship Differences in Male Ring Doves: Avoidance of Cuckoldry?" *Science*, 192:1353–1354.
60. Felsenstein, Joseph. 1974. "The Evolutionary Advantages of Recombination." *Genetics*, 78:737–756.
61. Foster, Mercedes S. 1977. "Odd Couples in Manakins." *Amer. Nat.*, 111:845–853.
62. Foster, Robin B. 1977. "*Tachigalia versicolor* Is a Suicidal Neotropical Tree." *Nature*, 268:624–626.
63. Fox, M. W. Editor, *The Wild Canids.* Van Nostrand Reinhold, 1975.
64. Fox, Laurel R. 1975. "Cannibalism in Natural Populations." *Ann. Rev. Ecol. Syst.*, 6:87–106.
65. Fricke, Hans, and Simone Fricke. 1977. "Monogamy and Sex Change by Aggressive Dominance in Coral Reef Fish." *Nature*, 266:830–832.
66. Geist, Valerius. *Mountain Sheep and Man in the Northern Wilds.* Cornell University Press, 1975.
67. Gheselin, Michael. 1969. "The Evolution of Hermaphroditism among Animals." *The Quarterly Review of Biology*, 44:189–208.
68. ———. *The Economy of Nature and the Evolution of Sex.* University of California Press, 1974.
69. Gilbert, Lawrence E. 1976. "Postmating Female Odor in *Heliconius* Butterflies: A Male-contributed Antiaphrodisiac?" *Science*, 193:419–420.
70. Gould, Stephen Jay. *Ontogeny and Phylogeny.* Harvard University Press, 1977.
71. Gould, S. J., and Niles Eldredge. 1977. "Punctuated Equilibria: the Tempo and Mode of Evolution Reconsidered." *Paleobiology,* 3(2):115–151.
72. Holldobler, Bert. 1976. "The Behavioral Ecology of Mating in Harvester Ants (*Hymenoptera*: *Formicidae*: *Pogonomyrmex*)." *Behav. Ecol. Sociobiol.*, 1:405–423.
73. Hogan-Warburg. 1966. "Social Behavior of the Ruff." *Ardea*, 54 (3,4):109–229.
74. Holm, Celia Haigh. "Breeding Sex Ratios, Territoriality,

and Reproductive Success in the Red-winged Blackbird (*Agelaius phoeniceus*)." *Ecology*, 54(2):356–365.

75. Horner, B. Elizabeth, and J. Mary Taylor. 1968. "Growth and Reproductive Behavior in the Southern Grasshopper Mouse." *J. Mamm.*, 49(4):644–660.
76. Howe, Marshall A. 1975. "Social Interactions in Flocks of Courting Wilson's Phalaropes (*Phalaropus tricolor*)." *The Condor*, 77:24–33.
77. Hrdy, Sarah Blaffer. 1974. "Male-male Competition and Infanticide among the Langurs of Abu, Rajasthan." *Folia Primatologica*, 22:19–58.
78. ———. 1976. "Care and Exploitation of Nonhuman Primate Infants by Conspecifics Other Than the Mother." *Advances in the Study of Behavior*. Vol. 6:101–158. Academic Press.
79. ———. 1977. "Infanticide as a Primate Reproductive Strategy." *American Scientist*, 65:40–49.
80. ———. 1977. *The Langurs of Abu*. Harvard University Press, 1977.
81. Hrdy, Sarah Blaffer and Daniel B. Hrdy. 1976. "Hierarchical Relations among Female Hanuman Langurs." *Science*, 193:913–915.
82. Huxley, Julian. *The Courtship Habits of the Great Crested Grebe*. 1914. (Cape Editions, 1968.)
83. Jacobs, Merle E. 1955. "Studies on Territorialism and Sexual Selection in Dragonflies." *Ecology*, 36:566–586.
84. Janzen, Daniel H. 1977. "What are Dandelions and Aphids?" *American Naturalist*, 111(979):586–589.
85. Jarman, P. J. 1974. "The Social Organization of Antelope in Relation to Their Ecology." *Behavior*, 58 (3,4):215–267.
86. Jenni, Donald A. 1974. "Evolution of Polyandry in Birds." *Amer. Zool.*, 14:129–144.
87. Jenni, D. A., and G. Collier. 1972. "Polyandry in the American Jacana (*Jacana spinosa*)." *Auk*, 89:743–789.
88. Johnsgard, P. A. 1968. *Waterfowl*. University of Nebraska Press, Lincoln, Nebraska.
89. Kessel, E. L. 1955. "The Mating Activities of Balloon Flies." *Systematic Zoology*, 4(3):97–104.

90. Ketterson, Ellen D. 1977. "Male Prairie Warbler Dies During Courtship." *The Auk*, 94(2):393.
91. Kleiman, Devra. 1977. "Monogamy in Mammals." *Quart. Rev. Biol.*, 52:39–69.
92. Krebs, J. R. 1969. "The Efficiency of Courtship Feeding in the Blue Tit *Parus caeruleus*." *Ibis*, 112:108–110.
93. ———. 1977. "The Significance of Song Repertories: The *Beau Geste* Hypothesis." *Anim. Behav.*, 25:475–478.
94. ———. 1977. "Communal Nesting in Birds." *Nature*, 269: 200.
95. Kroodsma, Donald. 1976. "Reproductive Development in a Female Songbird: Differential Stimulation by Quality of Male Song." *Science*, 192:574–575.
96. Kruijt, J. P., G. J. Vos, and I. Bossema. 1972. "The Arena System of Black Grouse." *Proc. XV Internat. Ornithol. Cong.*: 399–423.
97. Kruuk, Hans. *The Spotted Hyena*. University of Chicago Press, 1972.
98. Leonard, Jack, and Lee Ehrman. 1976. "Recognition and Sexual Selection in *Drosophila*: Classification, Quantification, and Identification." *Science*, 193:693–695.
99. Levi, H. W. 1976. "The Orb-Weaver *Genera Verrucosa, Acanthepeira, Wagneriana, Acacesia, Wixia, Scoloderus*, and *Alpaida* North of Mexico (*Araneae*: *Araneidae*)." Bulletin of the Museum of Comparative Zoology, 147(8): 351–391.
100. Luria, S. E. *36 Lectures in Biology*. M.I.T. Press, 1976.
101. Manning, J. T. 1976. "Is Sex Maintained to Facilitate or Minimise Mutational Advance?" *Heredity*, 36(3):351–357.
102. Mason, L. G. 1969. "Mating Selection in the California Oak Moth (*Lepidoptera, Dioptidae*)." *Evolution*, 23:55–58.
103. Mayr, Ernst. *Animal Species and Evolution*. Harvard University Press, 1965.
104. ———. *Evolution and the Diversity of Life*. Harvard University Press, 1976.
105. McCarty, Richard, and Charles H. Southwick. 1977. "Paternal Care and the Development of Behavior in the Southern Grasshopper Mouse, *Onychomys torridus*." *Behav. Biol.*, 19:476–490.

106. McKaye, Kenneth R. 1977. "Defense of a Predator's Young by a Herbivorous Fish: An Unusual Strategy." *Amer. Nat.*, 111:301–315.
107. McKaye, Kenneth R., and N. M. McKaye. 1977. "Communal Care and Kidnapping of Young by Parental Cichlids." *Evolution*, 31:674–681.
108. Nisbet, I. C. T. 1973. "Courtship-feeding, Egg-size, and Breeding Success in Common Terns." *Nature*, 241:141–142.
109. Metcalf, Robert A., John C. Marlin, and Gregory S. Whitt. 1975. "Low Levels of Genetic Heterozygosity in *Hymenoptera*." *Nature*, 257:792–794.
110. Mills, J. A. 1973. "The Influence of Age and Pair-bond on the Breeding Biology of the Red-billed Gull *Larus novaehollandiae scopulinus*." *J. Anim. Ecol.*, 42:147–161.
111. Milne, Lorus J., and Margery Milne. 1976. "The Social Behavior of Burying Beetles." *Scien. Am.*, 235(2) (August): 84–89.
112. Orians, G. H. 1969. "On the Evolution of Mating Systems in Birds and Mammals." *Amer. Nat.*, 103(934):589–603.
113. Osborne, David R., and Godfrey R. Bourne. 1977. "Breeding Behavior and Food Habits of the Wattled Jacana." *Condor*, 79:98–105.
114. Owen-Smith, Norman. "The Social Ethology of the White Rhinoceros." *Z. Tierpsychol.*, 38:337–384.
115. Parker, G. A., R. R. Baker, and V. G. F. Smith. 1972. "The Origin and Evolution of Gamete Dimorphism and the Male-female Phenomenon." *J. Theoretical Biol.*, 36:529–553.
116. Pietsch, Theodore W. 1976. "Dimorphism, Parasitism, and Sex: Reproductive Strategies among the Deepsea Ceratioid Anglerfishes." *Copeia*, (4):781–793.
117. Pitelka, Frank A., Richard T. Holmes, and Stephen F. McLean, Jr. 1974. "Ecology and Evolution of Social Organization in Arctic Sandpipers." *Amer. Zool.*, 14:185–204.
118. Purchon, R. D. *The Biology of the Mollusca*. Pergamon Press, 1968.
119. Ralls, Katherine. 1976. "Mammals in Which Females Are Larger Than Males." *Quart. Rev. Biol.*, 51:245–276.
120. ———. 1977. "Sexual Dimorphism in Mammals: Avian

Models and Unanswered Questions." *Amer. Nat.,* 111:917–936.

121. Rasa, O. Anne E. 1976. "Invalid Care in the Dwarf Mongoose." *Z. Tierpsychol.,* 42:337–342.
122. Reish, Donald J. 1957. "The Life History of the Polychaetous Annelid *Neanthes caudata* (delle Chiaje), Including a Summary of Development in the Family Nereidae." *Pacific Sci.,* 11:216–228.
123. Robel, R. J. 1967. "Significance of Booming Grounds of Greater Prairie Chickens." *Proc. Am. Philo. Soc.,* 111(2):109–114.
124. Robertson, D. R. 1972. "Social Control of Sex Reversal in a Coral-Reef Fish." *Science,* 177:1007–1009.
125. Sage, Richard D., and Robert K. Selander. 1975. "Trophic Radiation through Polymorphism in Cichlid Fishes." *Proc. Nat. Acad. Sci. USA,* 72(11):4669–4673.
126. von Schilcher, Florian, and Maurice Dow. 1977. "Courtship Behavior in Drosophila: Sexual Isolation or Sexual Selection?" *Z. Tierpsychol.,* 43:304–310.
127. Schuster, Richard H. 1976. "Lekking Behavior in *Kafue Lechwe.*" *Science,* 192:1240–1241.
128. Sherman, Paul W. 1977. "Nepotism and the Evolution of Alarm Calls." *Science,* 197:1246–1253.
129. Smith, J. Maynard. 1977. "Why the Genome Does Not Congeal." *Nature,* 268:693–696.
130. Snow, D. W. 1963. "The Evolution of Manakin Displays." *Proc. of the XIII Internat. Ornithol. Cong. (Ithaca)*:553-561.
131. Solbrig, Otto T. 1971. "The Population Biology of Dandelions." *Amer. Sci.,* 59(6):686–694.
132. Thornhill, Randy. 1976. "Sexual Selection and Paternal Investment in Insects." *Amer. Nat.,* 110:153–163.
133. ———. 1976. "Sexual Selection and Nuptial Feeding Behavior in *Bittacus apicalis (insecta: mecoptera).*" *Amer. Nat.,* 110:529–548.
134. Tinbergen, N. *The Herring Gull's World: a Study of the Social Behavior of Birds.* Collins, London, 1953.
135. ———. 1959. "Comparative Studies of the Behavior of Gulls (*Laridae*), a Progress Report." *Behavior,* 15(1,2):1–70.

136. ———. 1960. "The Evolution of Behavior in Gulls." *Scien. Am.*, 203(6) (December):118–130.
137. Trivers, R. L. "Parental Investment and Sexual Selection." In *Sexual Selection and the Descent of Man 1871–1971*, ed. by B. Campbell. Aldine, Chicago: 1972.
138. ———. 1976. "Sexual Selection and Resource-accruing Abilities in *Anolis garmani*." *Evolution,* 30(2):253–269.
139. Trivers, R. L., and Dan E. Willard. 1973. "Natural Selection of Parental Ability to Vary the Sex Ratio of Offspring." *Science,* 179:90–92.
140. Trivers, Robert L., and Hope Hare. 1976. "Haplodiploidy and the Evolution of Social Insects." *Science,* 191:249–263.
141. Van Den Ende, H. *Sexual Interactions in Plants.* Academic Press, 1976.
142. Van Engel, W. A. 1958. "The Blue Crab and its Fishery in Chesapeake Bay." *Commercial Fisheries Review,* 20(6): 6–17.
143. Vehrencamp, Sandra. 1977. "Relative Fecundity and Parental Effort in Communally Nesting Anis, *Crotophaga sulcirostris*." *Science,* 197:403–405.
144. Verner, Jared. 1964. "Evolution of Polygamy in the Long-billed Marsh Wren." *Evolution,* 18:252–261.
145. ———. 1965. "Breeding Biology of the Long-billed Marsh Wren." *Condor,* 67(1):6–30.
146. Verner, J., and Mary F. Willson. 1966. "The Influence of Habitats on Mating Systems of North American Passerine Birds." *Ecology,* 47(1):143–147.
147. Verner, Jared, and Gay H. Engelson. 1970. "Territories, Multiple Nest Building, and Polygyny in the Long-billed Marsh Wren." *The Auk,* 87:557–567.
148. Viljoen, S. 1975. "Aspects of the Ecology, Reproductive Physiology and Ethology of the Bush Squirrel *Paraxerus cepapi cepapi*." M. Sc. thesis, University of Pretoria.
149. ———. 1977. "Factors Affecting Breeding Synchronization in an African Bush Squirrel." *J. Reprod. Fert.,* 50:125–127.
150. Warner, Robert R., D. Ross Robertson, and Egbert G. Leigh, Jr. 1975. "Sex Change and Sexual Selection." *Science,* 190:633–638.

151. Warner, William. *Beautiful Swimmers.* Atlantic Monthly Press, 1976.
152. Wells, Kentwood D. 1977. "The Courtship of Frogs," from *The Reproductive Biology of Amphibians,* ed. by Douglas H. Taylor and Sheldon I. Gutman. Plenum Publishing Corp.
153. ———. "The Social Behavior of Anuran Amphibians." *Anim. Behav.* 25:666–693.
154. ———. 1977. "Territoriality and Male Mating Success in the Green Frog (*Rana clamitans*)." *Ecology,* 58:750–762.
155. ———. 1978. "Courtship and Parental Behavior in a Panamanian Poison-arrow Frog (*Dendrobates auratus*)." *Herpetologia,* 34(2):148–155.
156. White, M. J. D. "Heterozygosity and Genetic Polymorphism in Parthogenetic Animals." In: *Essays in Evolution and Genetics in Honor of Theodosius Dobzhansky,* ed. by M. K. Hecht and W. C. Steere. Appleton-Century-Crofts, NY, 1970.
157. Wickler, Wolfgang. *The Sexual Code.* Anchor Books, 1973.
158. Wilkinson, Paul F., and Christopher C. Shanki. 1976. "Butting-fight Mortality among Musk Oxen on Banks Island, Northwest Territories, Canada." *Anim. Behav.,* 24:756–758.
159. Williams, Ernest A., and A. Stanley Rand, 1977. "Species Recognition, Dewlap Function, and Faunal Size." *Amer. Zool.,* 17:261–270.
160. Williams, George C. *Sex and Evolution.* Princeton University Press, 1975.
161. Wilson, E. O. *The Insect Societies.* Harvard University Press, 1971.
162. ———. *Sociobiology.* Harvard University Press, 1975.
163. ———. Editor, *Ecology, Evolution, and Population Biology.* W. H. Freeman, 1974.
164. Wooton, R. J. *The Biology of the Sticklebacks.* Academic Press, 1976.
165. Worden, A. N., and J. S. Leahy. 1962. "The Behavior of Rabbits." In *The Behavior of Domestic Animals,* ed. E. S. E. Hafez. Baillière, Tindall, Cox. London.
166. Zahavi. 1975. *J. Theor. Biol.,* 53:205–214.

INDEX

adaptive elite theory, 120–127, 146
aggressive behavior, 182
  in asexual vs. sexual systems, 62
  by females, 159, 165–166
  infanticide as, 129–133, 136–138, 166
  male competition and, 107–109, 114, 136–138, 142–143
alarmone, 54–55
algae, 35, 36, 38–39, 51, 59
altruism, 54–56, 60–63, 65–66, 96
  definition of, 60–61
  kinship and, 62–65
amoebas, 35*n,* 54–55, 60–61
amphibians, 41, 100, 113*n,* 157–158
anglerfish 73–75, 152, 154*n*
anis, groove-billed, 59–60
antelope, African, 100, 104–105, 153
ants, 62, 63
  harvester, 118–119, 120
aphids, 43*n,* 63
Armitage, Kenneth, 101–102

baboons, 112, 141
bacteria
  asexual reproduction of, 28–29, 30, 185
  boom-or-bust population style of, 27, 46
  description of, 26–27
  as dominant life-form, 25
  evolution of, 26–27
  gene characteristics of, 28–30
  habitats and, 46, 47
  low-energy states of, 28, 46
  protozoa and algae vs., 35
  reproduction rates of, 28–29
  sexual reproduction of, 29
  in symbiotic relationships, 59
Barash, David, 180–182
barnacles, 76, 152
Bartholomew, George, 115–117
Beautiful Swimmers (Warner), 147–148
beavers, male parenting by, 159–160
bees, sand, 97–99, 112
beetles, burying, 154–155, 175
birds
  cooperation-competition links and, 59–60, 103
  courtship displays of, 89, 92, 103–104, 114, 127
  duetting among, 184
  egg wastage among, 59–60
  female accommodation in, 141–146
  female competition among, 165–166
  female incitement in, 114–115
  fight-or-flight reflexes in, 93–95

birds (*cont.*)
helpers-at-the-nest phenomenon in, 63–65
male parenting among, 16, 154, 158–159, 167–168, 175, 181–182
monogamy among, 133–134, 168–169, 181–185
opportunistic breeders among, 185
parasitic, 153–154
size in, 113*n*
vulnerability of young of, 177–178
birds-of-paradise, 167, 185
bisexuality, 32, 186
advantages of, 71
defined, 31
distribution of, 75, 76
explanation for, 76
bluebirds, mountain, 181–182
bonito, Pacific, 114
bowerbirds, New Guinea, 18, 167
Brachionis, 41
butterflies, 22

Calder, William, 170–171
canaries, 92
cannibalism, 152, 154*n,* 166
cats, 92
Chlamydomonas reinhardi, 38–39
chronological elite, 146–150
Cichlasoma, 50
Cichlidae, male parenting by, 155–157, 169, 175
cloning, *see* reproduction, asexual
clovers, 59
clown fish, 81–83
Clutton-Brock, Tim, 108–109
cockroaches, 36–37, 38
Collias, Nicholas and Elsie, 142–146
colobines (leaf-eaters), 136–137
competition, 17, 22, 23–24, 45, 66, 67, 96–109, 152, 154*n,* 174
altruism vs., 61–63
cost to females of, 128–138
in Darwin's theory, 21–23, 61
female control of, 110–127, 134–136
female-female, 77, 145–146, 150–151, 159, 163–167
forms of, 17–18
gene sets and, 117–118, 119
kin selection theory and, 62–65
limitations on, 111, 123, 139–150
male aggregations in, 103–105, 106
maturation delays and, 106
mortal battles in, 107–109, 137–138, 142–143
paternal care and, 168, 172–173
sexual reproduction and, 51–52, 59, 61–62, 66, 67–68, 70, 72, 75, 76–77, 91–92, 95
size and, 45–47, 51–52, 113–114, 120, 139, 153, 169, 179
social, 45–47, 50, 66, 103–105, 106, 177–179, 186–187
specialization and, 48, 50–52, 67, 179, 185–186
vulnerability of young and, 177–178, 179
see also territorial rights
cooperation

in asexual vs. sexual systems, 58–59, 66
competition linked with, 59–60, 66, 68, 103–105, 106, 174
kin selection theory and, 62–65
see also organization; parenting, male
copepoda, 75–76
copulation
dominance as goal of, 115–117
mechanical stimulation of, 92–93
coral fish, 81–83
Coulson, J. C., 183–184
courtship displays, 18, 144, 147, 150, 162–163, 164–165, 168
aesthetic function of, 127
as communication, 86–90, 164
as competition, 97, 99, 153
endocrine changes and, 92–95
evolution and, 170
female feeding and, 170–172
fight-or-flight reflexes and, 93–95, 106
threats and, 114, 142–143
cow dung, 48
cowbirds, 153
Cox, Cathleen, 117, 121–123, 140
coyotes, American, 112, 124
crabs
Chesapeake Bay blue, 146–150
fiddler, 86, 155
hermit, 40
cuckoos, 153

Darwinism, 16, 20–23, 25, 33, 59, 76, 79, 109
Davies, N. B., 113–114
deer, red, 108–109
Dillard, Annie, 15
dinosaurs, 109
DNA, in protista, 36
dogs, wild, 17–18, 160, 166–167
doves, ring, 167, 182
dragonflies, 100, 128
ducks, 114–115, 133
duetting birdsong, 184
dwarf-male system, 152

earthworms, 72
ecology
dewlap patterns and, 90
excess fecundity and, 45–52
exclusion principle in, 48–49
gender flexibility and, 78
male parenting and, 177
see also habitat
eels, American, 75
epuresis (enurination), 88
estrus cycle, 70, 110, 111, 112, 114, 120, 138, 140, 141
in synchrony with males, 161, 163
evolution
altruism vs. competition in, 61–62
courtship and, 170
dependency period and, 178
Eastern vs. Western theory of, 27
female choice in, 170–172
of gender, 67–77, 186
toward hermaphroditic species, 186
masculinity as alternative in, 23–24
monogamy and, 182–187
sex as antisocial force in (Wilson), 54

evolution (*cont.*)
  sexual maturation schedules and, 178
  sexual reproduction selected in, 34, 44, 52, 67–68
  specialization and, 16, 44, 47 52, 186
  survival theory of, 22–23
  see also Darwinism
exclusion principle, 48–49
extraterrestrial bacterial spores, 28

female feeding, 170–172, 173
fertilization
  "chronological elite" and, 146–150
  laboratory-induced, 72
fight-or-flight reflexes, 93–95, 106
fireflies, 17, 70, 103
fish
  altruism in, 156–157
  asexual reproduction of, 41
  cooperative assemblies of, 58, 103
  courtship displays of, 90, 162–163, 164–165
  gender flexibility in, 79–82
  male parenting and, 21, 155–157, 161–163
  maturation delays in, 106
  parasitic males in, 74, 152
  in reproductive cycle, 21, 69–70, 72, 73–74, 99–100
  size of, 113*n*
  territorial competition by, 100
flatworms, 43
flies, 42, 48–49, 50, 70, 85, 90–91, 106
  balloon, 134
  eastern scorpion, 171–174, 180
flocking, 58
food chain, 26
foraging, breeding vs., 69, 74, 76, 84, 153
Forel, Auguste, 63
Foster, Mercedes, 93–94
foxes, 178
Fricke, Hans and Simone, 82
frogs, 72, 85, 92, 103, 106
  neotropical, 152, 154*n*, 157–158, 166–167
fruit flies, 48–49, 90–91
fungi, 42, 54–56, 59

gall midges, 42–43
Galton, Francis, 58
gazelles, 17–18
Geist, Valerius, 17
gender
  defined, 30–31
  determination of, 78, 83, 134–135
  evolution of, 67–77, 186
  flexibility of, 78–83
  gene transfers and, 34
  "+" and "–" as, 38
  as specialization, 50, 77, 179, 185–186
gene sets, 117–118, 119–120
gene-copying, in bacteria, 29
gene-labeling hypothesis, 89–90
genetic-elite theory, 120–127, 146
gibbons, 184
Gier, H. T. 112, 124
Gilbert, John, 42
giraffes, 70
goldfish, 72
grasshoppers
  American, 18
  European, 89–90
great chain of being, 178
gulls, 175

kittiwake, 183, 187
red-billed, 184

habitat
asexual vs. sexual reproduction and, 39–40, 42–44, 51
breeding systems and, 186
spanadry and, 75–76
see also ecology; territorial rights
hadrosaurs, 109
Halliday, T. R., 113–114
"handicap principle" (Zahavi), 124, 127
Hare, Hope, 62–63, 135
heath hens, 104
helpers-at-the-nest phenomenon, 63–65
Hölldobler, Bert, 118–119
Hrdy, Sarah Blaffer, 129–133, 137, 138
hyenas, 138

impalas, 100
incitement
benefits to males of, 141
in female competition, 165
in male competition, 112–115, 120, 121–127, 136
individuation
altruistic sacrifice and, 54–56, 60–63
evolutionary, 44, 47, 52, 66
see also organization
infanticide, 129–133, 136–138 166
information exchange, 68–70, 84–91, 95, 108, 164, 167, 184–185
insects, 49, 50, 75–76, 85, 100, 103, 105, 119. 134, 152, 154*n*
male parenting by, 154–155
see also social insects
invertebrates
gender flexibility of, 78–79
male parenting by, 154–155
males from unfertilized eggs among, 119
see also marine invertebrates

jacanas, American, 158–159
jackals, 160
Jacobs, Merle, 128, 151
jellyfish, 53–54, 59
Jenni, Donald, 159
Jimmy potting, 149
Johnsgard, Paul, 114–115

Kessel, Edward, 134
kin selection theory, 62–65

langurs, hanuman, infanticide among, 129–133, 136, 138, 166
Le Boeuf, Burney J., 117, 121–123, 140
lek systems, 103–105, 106, 118, 127
male advantages in, 139–140
leopards, 133, 136
lichens, 59
life, defined, 15
life span, competitive mating and, 17–18, 22, 109
Linnaeus, 27*n*
lions, 100–101, 111, 133
living things
dead things vs., 15
distinctions between, 35, 41
lizards, 141
asexual, 43–44
copulating behavior of, 120–121
courtship displays of, 87, 88–89, 90, 92

lordosis, 93
love, meaning of, 187
low-energy states, 28, 46

McKaye, Kenneth, 155–157
males
  asexually vs. sexually produced, 57, 119
  compensating advantages of, 23–24, 33–34, 71, 76–77, 174–180, 185–187
  discrimination reduced in, 85–86
  in evolution of gender, 23–24, 67–77, 179, 186
  female accommodation to, 140–150
  female control of competition among, 110–127
  female dependency on, 179–187
  in female's service, 71–77
  gender flexibility and, 78–83
  in kin selection theory, 62–63
  with one vs. two gene sets, 117–118
  population control of, 70, 75–76, 91, 134–136, 139, 151, 161, 164
  reproducing vs. peripheral, 140
  see also parenting, male; reproductive cycle, male; specialization
males vs. females
  balance of power between, 139–187
  as best evolutionary alternative, 23–24
  dependency needs and, 179–187
  function in nature of, 18–19
  harshness of life and, 17
  in nonmating interactions, 139
  size of, 113*n*, 120, 138, 139
mallards, 133, 180–181, 182
mammals, 167
  courtship behavior in, 92, 103
  duetting by primates among, 184
  lek system in, 104–105
  male parenting by, 159–161, 175
  male vs. female size of, 113*n*
  vulnerability of young of, 177–178
manakins, 93–95
man-of-war, 53
mantids, 152
marine invertebrates, 75–76, 155, 165, 166
  reproduction system of, 69, 72–73, 79
marmosets, 160
marmots, yellow-bellied, territorial competition of, 101–103
marsh birds, 141, 165, 176
mate-choice experiments, 168
mating
  conditions for, 68–69, 84–95
  noise in, 17
  quickness of, 18, 85
  synchronization in, 91–92
mating plugs, 105–106
means-of-population-control spectrum, 44–52
mice, 72
  deer, 160, 168
  desert, 160–161
millipedes, 76
Milne, Lorus and Margery, 153
mimetic (sibling) species, 89–90
minnows, 72
mites, 119

mollusks, 41, 78
mongooses, altruistic behavior in, 65–66
monkeys, 129–138, 160
monogamy
  control reversal and, 152–153
  efficiency of, 185
  gender leverage in, 168–170
  hybridization prevented by, 185
  male parenting and, 167–168, 181
  among mammals, 65
  of parasitic birds, 153–154
  positive vs. negative aspects of, 181–187
moths, 22, 69
mountain sheep, 17
mouthbreeding fish, 21
multicelled organisms, sexual options of, 41–44
musk-ox, 107–108

Neetroplus nematopus, 157
nematodes, 48, 72
nesting
  communal, 59–60
  in male-female controls, 141–146
  of parasitic birds, 153–154

orangutangs, 133
orchids, 78, 85–86
organization
  asexual reproduction and, 57–58
  dependency in, 58–59, 66, 178–179
  of females, 110, 111–112, 120
  gender selection by females and, 134–135
  hierarchies in, 60, 65, 115–117
  individuality and, 58–60
  specialization and, 54, 56, 59
  see also lek systems
Origin of Species, The (Darwin), 20
ovipositing, 100

pair bond, *see* monogamy
paramecia, 35*n,* 40
parasites, 78, 135, 173
parenting, male, 16, 23, 154–161, 165, 166, 167–168, 169, 174
  ecological factors and, 177
  evolution of, 170–180, 186
  experiments with, 180–182
  female responses to, 176, 179
  offspring dependency and, 177–178, 179, 186
  risks in, 173–174, 180–181
peacocks, 127, 167
penguins, emperor, 165, 167
phalaropes, Wilson's, 165–166
pheromones, 69, 74, 91, 99, 114, 119, 147
  antiattractant, 106
  as food signals, 154
  in territorial control, 103
  as trap bait, 168
pigeons, 60, 167, 168, 169
pipefish, 155, 166
planarians, 72
plant-animal system, five kingdom organizaiton of, 27*n*
platyfish/guppy, 106
Polyspira, 40
population pressures, 70, 75–76, 91, 134–136, 139
  bisexuality and, 75
  male parenting and, 177, 179
  means-of-population-control and, 44–52

population pressures (*cont.*)
  reproductive changes and, 42, 43, 44
Pouyanne, A., 86
prairie chickens, greater, 140
prairie dogs, 101
primates, arboreal, 175
progeny
  diversification of, 34, 44, 50–52, 67
  safeguards for, 177, 186
progesterone, 93
protista
  asexual vs. sexual reproduction of, 36–41
  bacteria vs., 35, 36
  description of, 35–36
protozoa, 35, 36–38, 40
pseudotransvestism, 106, 173

quails, button, 166

rabbits, 87–88, 92
rape, 133, 180–181
Rasa, Anne, 65–66
rats, 92–93
reindeer, 125
reproduction
  excess fecundity and, 44–52
  genes and, 34, 66, 72, 117–118, 119–120
  homeostatic equilibrium in, 135
  at juvenile stage, 43*n*
  promoted by evolution, 22, 34, 44, 52, 67–68
  rates of, 28–29, 42, 43*n*
  scientific terms for, 40–41
  sexual options in, 25, 30–31, 37–44, 68
reproduction, asexual, 25, 32, 61
  of bacteria, 28–29, 30, 46, 185
  defined, 30, 32
  as dominant system, 25, 33
  of multicelled organisms, 41–44
  of protista, 36–41
  size of organism and, 28, 33, 43
  specialization and, 54, 185
reproduction, sexual, 25
  as adaptive, 67
  aggressive behavior and, 62
  competitiveness promoted by, *see* competition
  consequences of, 53–66, 182–187
  defined, 30
  evolutionary selection of, 34, 44, 52, 67–68
  laboratory induction of, 39, 42
  location and, 84, 140, 145–146, 147, 153
  as low-fidelity system, 33
  mutuality decreased in, 61, 66
  prefertilization activity and, 69, 71–72, 84–95, 142
  pros and cons of, 30, 32–34
  schedules of, 71–72, 74, 84, 91, 110, 120, 140, 146–150
  size of organism and, 43, 51–52
  as social adaptation, 51–52
  specialization and, 50, 51–52, 59, 66, 67, 68, 77
  synchrony in, 161, 163
  transition to, 35–52
reproductive cycle, female
  choice of mate in, 70, 75, 84–91, 95, 112–113, 117–127, 139, 150, 163, 164–165, 168, 176–177
  control and, 16, 71, 74–75, 96, 110–127

dangers in, 85, 91
efficiency of, 71–72, 77
endocrine changes and, 92–95
initiatives in, 69–70, 73, 112–115, 120, 121–127, 136, 141, 148–149, 150, 159, 162–163, 165–167, 173
male parenting and, 176, 179–180
polygamy as advantageous to, 176–177

reproductive cycle, male
choice of mate in, 145, 146, 150–151, 163–164, 165–167, 168, 169
competition in, *see* competition
control in, 139–150
courtship in, *see* courtship displays
as critical, 174–176
dangers in, 17–18, 22, 70, 107–109, 137–138, 142–143, 152, 173–174, 180–181
dwarf-male system in, 152
female feeding and, 170–172, 173
leverage and equality in, 151–187
as peripheral role, 16, 70
sperm and, 119–120, 123
see also males; parenting, male

reptiles, 41, 92, 100, 105, 113*n*, 157
rheas, American, 180
roadrunners, 170–171
Robertson, D. R., 80
robins, Australian, 89
rodents, 105, 119, 159–161
rotifers, 41–42, 43
ruffs, 103–104, 127

sables, 100
salamanders, 72
salmon, 69–70, 99–100
scale insects, 49, 50
Schelling, F. W. J. von, 178
Schuster, Richard H., 104–105
scrub jays, 64–65
sea anemones, 43, 81
seahorses, 155, 166
seals, elephant, 85, 109, 128–129
adaptive elite theory and, 121–123
female incitement and, 115–117
male competition and, 111–112

sexual reproduction, *see* reproduction, sexual
Shank, Christopher, 108
Shaw, George Bernard, 33
shelducks, 114–115
Sherman, Paul, 64
shore birds, 127, 158–159
"shouting," olfactory, 102
shrews
Australian marsupial, 152
elephant, 176

siamangs, 184
single-celled organisms, in life-form divisions, 35, 41
Siphonophora, 53–54, 59
size
asexual reproduction and, 28, 33, 43
boom-and-bust population cycles and, 27, 46
competition and, *see* competition
of large vs. small creatures, in ecology, 47, 51–52
sexual reproduction and, 43, 51–52

skuas, Arctic, 184
slime mold, 54–56, 60–61
slugs, 75
Smith, Hugh M., 103
snails, 72, 75, 79
snakes, 64
  garter, 99, 112
"sneaky fucker strategy"
    (Clutton-Brock), 109
social environments
  distribution of organisms by,
    185–186
  gender flexibility and, 78–83
  specialization and, 47, 50
  see also competition
social insects
  altruism of, 56
  as asexual, 57–58
  gene sets of, 63*n*, 118
  kin selection theory and,
    62–63
  male-female ratios in, 135
  organization of, 56–57
Sociobiology (Wilson), 54
spanadry, 75–76
specialization
  competition as force for, 48,
    179
  disadvantages of, 50–51
  evolutionary, 16, 44, 47, 52,
    186
  exclusion principle and, 48–49
  gender, 50, 77, 179, 185–186
  monogamy and, 184–187
  in organizations, 54, 56, 66
  sexual reproduction and, 50,
    51–52, 59, 66, 67, 68, 77
sperm
  supplies of, 123
  survival of, 119–120
spiders 17, 113*n*., 115, 152
sponges, 75, 78
spores, bacterial, 28
squirrels, 64
  African bush, 114
starlings, 178–179
sticklebacks, 161–165, 175, 180
  reproductive cycles favoring
    males of, 161–163
stress tests, 117–120, 144, 165
Sugiyama, 130
sunfish, 90
swagger matches, 107
symbioses, 58–59, 179–180
symbolic conflict, 107–109, 114
syndesmogamy, 40
syzygies (pairs), 40

Tacitus, 107*n*
talking birds, 184
tamarins, 160
teals, green-winged 133
territorial rights, 100–103, 104-
    105, 142, 145–146, 159,
    176–177
Thornhill, Randy, 171–172, 173
tits, blue, 170
toads, 85, 113
tools, adaptation and, 68
tortoises, 87
traits, inheritable, 84
  adaptive elite theory and, 120
  aesthetic criteria and, 126–
    127
  altruistic, 61–63
  in Darwinism, 20, 21–22
Trichonympha, 36–38
Trivers, Robert, 62–63, 121,
    125, 135, 166
turkeys, wild, 106

Vehrencamp, Sandra, 59–60
vertebrates
  asexual, 43
  gender flexibility and, 78–83
Viruses, 25*n*

Warner, William, 147–148, 149
wasps, 17, 86
waterbucks, 100
water fowls, hole-nesting, 169*n*
water striders, 87
weaverbirds, female accommodation among, 141–146
Wells, Kentwood, 157–158, 166–167
whales, 113*n*
White, T. H., 57
Wickler, Wolfgang, 37
Wilkinson, Paul, 108
Wilson, E. O., 54, 56–57, 154–155, 184
wolves, 160
Woolfenden, 63–64
Wooten, R. J., 162
worms, 43, 48, 75, 78, 105, 155, 165, 166
wrasses
  bluehead, 79–80
  cleaner, 80–81

Xylocaris maculipennis, 105

Zahavi, Amotz, 124, 127
zygopalintomy, 40–41

ABOUT THE AUTHOR

Fred Hapgood is a professional science writer. For five years he was the science reporter for the Harvard News Office, where he covered both the research and the controversy generated by the new science of sociobiology. His work has appeared in several national magazines including the *Atlantic Monthly*, and he is the author of one other book, *Space Shots: A Picture Album of the Universe.*